AF553197

Fish and Fisheries of India

About Author

Dr. Sandhya S. Kadam (b. 1955), obtained her M.Sc. Ph.D. (Zoology) from Dr. Babasaheb Ambedkar Marathwada University, Aurangabad, Maharashtra, India. Presently she is working as Associate Professor and In Charge - Principal, Dnyanopasak College, Parbhani, Maharashtra, India. She has to her credit about 20 research papers published in Journals and in books of national and international repute. One book entitled "Hydrobiology of the Reservoir: Enhancement in Capacity and Assessment of Quality" is published by Lambart Academic Publication Germany and second book entitled "Aquaculture and Fish Technology: Fundamentals, Applications and Economics" is under to publish by Lambart Academic Publication Germany in the month of August 2016. She is the Secretary of the Dnyanopasak Shikshan Mandal, Parbhani India. She worked on the Bhogaon Reservoir from Parbhani District Maharashtra from the point of view of desilting, deweeding and hydrobiology and created the Bhogaon Pattern.

Fish and Fisheries of India

By

Dr. Sandhya S. Kadam
In-charge Principal and Head
Department of Fishery Science
Dnyanopasak College, Parbhani (Maharashtra)

NEW INDIA PUBLISHING AGENCY
Pitam Pura, New Delhi – 110 088

NEW INDIA PUBLISHING AGENCY
101, Vikas Surya Plaza, CU Block, LSC Market
Pitam Pura, New Delhi 110 034, India
Phone: + 91 (11)27 34 17 17 Fax: + 91(11) 27 34 16 16
Email: info@nipabooks.com
Web: www.nipabooks.com

Feedback at feedbacks@nipabooks.com

ISBN: 978-93-85516-83-2

Composed, Designed and Printed in India

Preface

Considering the need of suitable text book which may fulfill the requirement of the students, teachers and researchers I felt to bring this book on "Fish and Fisheries of India". Having an experience of nearly two decades to teaching fishery science, the present edition of the book on Fish and Fisheries of India will accomplish the long felt need of the students. The effort has been made to keep language easy. The topics have been selected according to the University Syllabi. The target audience of the book is the students, teachers, researchers of disciplines of life sciences (Zoology and Fishery Science).

The data and inferences presented in the book have been taken from the various sources. The matter has been explained with the help of finely drawn, simple and well labelled diagrams, where ever necessary, tables and charts have been given for clear understanding of the subject. Author is grateful to colleagues, students and friends for the valuable suggestions and help in preparation of this book. Author is also grateful to the publishers for bringing the best possible book. This book focuses on the various aspects of fish and fisheries of India. The contents have been organized in 'self-contained' units for ease of understanding. The entire book is divided into 20 chapters.

Confronted within the limitations and the tedious task as compared to the time and other busy assignments, the author has every right to confess at the outset that they might have fallen prey to some omissions or some prejudices, which could have introduced discrepancies in the treatment and sometimes even in outcomes and would have denied the work a sort of perfection, for which they have a genuine craze for all of which they would like to be excused.

At the outset I humbly submit my feelings of gratitude towards venerated Adv. Ganeshraoji Dudhgaonkar (Bapusaheb) President, Dnyanopasak Shikshan Mandal Parbhani (Maharashtra) for being the source of inspiration for bringing out this book. I am also indebted to Dr. S.L. Sadawarde (Principal), and Dr. Sunil Kadam Department of Fishery Science, Dnyanopasak College Jintur and Dr. Md. Babar Department of Geology, Dnyanopasak College, Parbhani, who gave all the possible help and support for preparation of the book. I would like to express my indebtedness for the affection and blessings from my family members, which had been and will remain a constant source of inspiration throughout my life.

—Dr. Sandhya S. Kadam

Contents

1

History of Fishery Science, Scope and Importance

A fish is defined as the cold blooded vertebrates, which catch with the help of gills and swims with the help of fins and which are aquatic in nature. In broader sense the term fishery is used for the rearing and management of all aquatic animals which are useful to human being e.g. fishes, crustaceans and mollusks. The term fishery science is related to the rearing and management of fish on large scale by proper utilization of available water sources.

Fishes comprise of about 33100 species (FishBase 2015) differing widely from each other in shape, size, habits and habitats. Some of them are very small or more than an inch in length, while a few attain a length up to 18.50 m. they live in all free flowing rivers, lake, canals, dams and in almost every place where there is water.

Fishes usually have a stream lined body but some are elongated snake like and a few are dorso-ventrally compressed. They have paired and unpaired fins supported by soft or spiny fin rays. Dorsal, caudal and the anal fins are unpaired while the pectorals and pelvic (ventral) are paired fins. A number of species possess barbless which are excellent organs of touch.

They inhabit a variety of different kinds of environments ranging from deep water of the ocean to the boundless surface of the open sea and from high tropics to the receding Polar Regions; some are live in muddy water of the bays brackish estuaries stagnant pools and in the water several feet under the ground in caves. Of these 41.2% inhabit the fresh water and 58.2% the marine water of the world. The fresh water of the world constitute only 0.0093% as against the 97% of marine water. It is significant to include that the number of fresh water species is greater than marine species.

Table 1 : Total number of fish species living in different habitats.

Sr. No.	Habitat	Percentage
1.	Fresh water species	41.2
2.	Diadromus fish species	0.6
3.	Marine, shore and continental shelf to 200 m in warm water	39.9
4.	Marine, shore and continental shelf to 200 m in cold water	5.6
5.	Deep sea, benthic below 200 m	6.4
6.	Epipelagic above 200 m	1.3
7.	Deep sea pelagic below 200 m	5.0

It appears that in two kinds of environments the mode of formation and the structure of niche are highly divergent. Wide distribution of fishes has also resulted into two production of a variety of different kinds of adaptive modifications in order to meet with the stresses of the physical chemical and biotic environments.

General Characteristics of Fishes

All the fishes are built on a fundamental plan and process following characters in general.

1. They are cold blooded i.e. poikilothermic vertebrates and then body is suited to aquatic mode of life.
2. The locomotion occur by paired (pectoral and pelvic fins) and unpaired (dorsal, anat and caudal fins) fins. Fin also help in maintaining equilibrium of the body. They are supported by true dermal fin rays.
3. The body is covered by dermal scales.
4. The notochord is partially replaced by formation of vertebrae which are either cartilaginous or bony in nature.
5. Skin and visceral arches are well developed paired visceral arches are present, the 1st jaws forms the upper and lower jaws the 2nd forms suspensorium and rest support the gills.
6. Respiration occurs through gills.
7. Heart is venous and the circulation of single type i.e. only the venous blood flows through the heart.
8. The kidney of adult is mesonephros, although a non-excretory pronephros may persist in adults.

9. Allantoic bladder is absent.
10. Lateral line sense organs are well developed and housed in lateral line canals.
11. A swim bladder is usually present (except in elasmobranchii) but is secondarily lost in some species.
12. The nerves are mostly medullated.
13. The dorsal and ventral roots of spinal nerves are united at base.
14. A middle ear is absent and the internal ear is provided with the usual three semicircular canals.

Among the fish consuming countries, Japan, exploits most of the surrounding seas feeding with fish then may either country. According to FAO statistics countries may be classified on the basis of their per capita consumption of fish as follow:

1. Very high i.e. 20kg or more : Iceland, Norway, Sweden, Burma, Japan and Philippines .
2. High i.e. 10kg -20kg. : Belgium, Denmark, England, Holland, Portugal, Germany, Spain, S. Korea, Siam, and central U. S. Countries.
3. Medium i.e. form 5 kg.- 10 kg. : Finland, Greece, France, Italy, Luxembourg, Canada, USA, Chile, Venezuela and Australia.
4. Low as less than 5 kg : Ireland, Austria, Yugoslavia, Mexico, Switzerland, India, Turkey, Central and South America.

The evolution of earliest vertebrates occurred in marine environments. The long diversified history of vertebrates begins with the fragment of vertebrate bone, which for the 1st time appeared in the fossil records of Ordovician.

The most ancient vertebrates called agnatha or jawless vertebrates. *Agnathas* had undergone an extrusive adaptive radiation during the Devonian and gave rise to a variety of different kinds of armored plated fishes. These fishes are called as the *Ostracoderms*.

In this evolution periods the 1st pteraspldomophs (disparhina = two names) were among these *Ostracoderms* to be appear in the fossil record. Then the *Cephalaspidomorps* (monorhina = one nare) which were represented by three different groups of fishes is *Osteostracl, Anaspid* and *cyclustone*. Of these only *cyclostomes* are survive and latter two becomes extinct on carboniferous nearly 300 million years.

In vertebrate evolution the sudden appearance of ganathostomes in the mid-Silurian period. These fishes underwent a great adaptive radiation and gave rise to many kinds of the fishes. The fishes of these kinds are called acanthodes and another groups these are called placoderms.

Placoderms, however could not survive for long and disappeared gradually as two new types, the cartilaginous and bony fishes evolved in late Silurian and early Devonian period. Thus forms jawless to jawed animals are to be evolved in the succession period of evolution.

In evolution period from ostracadevest which are first appeared and then modified and developed in continuous succession and forms higher developed fish is teleostomii, these are well developed fish class.

Scope of Fishery Science

Fishery science is a super science. It has several constituent basic sciences drown into its fold from several different disciplines.

Modern fishery science has emerged only as a result of a very beautiful blending of the many basic sciences such as ecology, biology, hydrology, meteorology, astronomy, biochemistry, microbiology, pathology, economics and so on. One common goal of these all is "the fish as food for mankind". The promise and problems in the fishery science are perhaps never-ending. Nevertheless, the progress made in the last few decades has been tremendous. We have begun to understand our limitation in the exploitation of the available natural resources of fish.

The practical uses of our knowledge gathered so for in the field in themselve are encouraging and shall serve as a guiding force for the new generation need to establish protected water for fish in general and sanctuaries for others in particular is one such revelation. Experimental intensive cultural in laboratory could give yield of fincarps weighing 9130 gms in a 40 lit capacity aquarium using closed recirculation system.

The herbivorous fish gives yield ten times greater than that given by a carnivorous fish on an average. Apart from the food importance there is one more important consideration of the fishery industry. It plays an important role in the rural economy. The immense job potential it can provide can go a long way to ease the pressure of unemployment currently engulfing many parts of world.

Some of the special fisheries like those of sardines and mackerels support the living of several thousand fishermen in India. There is equally great scope of self employment in fish culture industry in rural areas.

Generally in recent days fish production is increased. As population increased the food problems create to solve the food problems fish culture is most important and it is very easy.

In modern days Biotechnology is introduced using these techniques. We increase the fish species as our water sources. Then we cultured the fish as disease resistance using genetic engineering processes. Large number and large amount or quantity of fish is destroyed by different action to decrease such loss human being performs different works settle the self industry and product made.

Different valuable by products are prepared and these are sold in to international market it is again a best method. Where there is water is present in such water on a viewpoint that fish culture is occurred seed production improves the different variety of fishes which live in local area. In these days it is the best source for job work as well as other different by products are manufactured and these are sold and get more money.

Importance of Fishery Science

Fishes forms one of the most important groups of vertebrates for man, influencing his life in various ways millions of human beings suffer due to hunger and malnutrition. It has following importance.

i. Food – Fish is an excellent food for mankind is more easily perishable than cattle, sheep and chicken.

ii. Fish flesh has very great nutritive and medicinal values.

iii. Fish flesh is easily digestible by child and man.

iv. Fish food provides proteins and vitamins required for the body of human being.

v. All the essential amino acids are present in the fish flesh.

vi. Fish flesh contains proteins, fats, vitamins, waters and minerals.

vii. Fish flesh contain about 13-20% proteins. It is easily digestible and nutritional.

viii. Fish proteins are formed of all 10-11 essential amino acids i.e. lysine, valine, arginine, itistidine, isotericine leucine, metholinine, cyltine, phenylalanine, tyrosine, threonine tryptophan etc.

ix. Fish flesh contains about 1-18 fats or oils, The lipid content of the fish flesh is also high but it varies in different species of fishes according to the environment.

x. The fish oils are the most important value of fishes. These oils are combines with vit, sterols, hydrocarbons phospholipids and coloured substances as squalene.

xi. Flesh of fish is highly perishable- (liable to spoil or decay) commodity (Article of trade) constituted by 60-80 % water.

xii. Some minerals are present in flesh of the fishes these are calcium, potassium, phosphates, Iron, copper, zinc, magnesium and Iodine.

xiii. Fish flesh contains a very high percentage of vitamins such as A, D, E and K.

xiv. Certain by- product of fish industry are used in Ayurvedic and Unani system of medicine.

xv. Certain by products are used in night- blindness, skin diseases, colds cough, bronchitis, asthma and tuberculosis.

xvi. Fish is used as the food of cattle and poultry is called fish meal which is very high nutritive value.

xvii. Fish that are unfit for human consumption are used to prepare fish manure for the fields.

xviii. From the fish, the fish glue and isinglass is prepared.

xix. The skin of several fishes like the sharks and rays are used for preparing ladies shoes, money bags, suitcases belts etc.

xx. Some fishes are used for biological control such as malaria.

xxi. Fish is useful for manufacturing the Artificial pearl.

xxii.. Fish is also useful for keeping in aquarium as a hobby.

2

Systematic Position of Pisces among the Chordates up to Class Level

What are Chordates?

The animal kingdom is divided first into several major animal groups called phyla: singular phylum. There are approximately so animal phyla. The last major groups of the animal kingdom are known as phylum chordate it was created by Belfour, in 1880. The name of this phylum is derived from two Greek words, the chorda- meaning a string or cord and ata- meaning bearing. The reference is to a common characteristic feature in the form to a stiff, supporting rod – like structure along the back, the notochord (Gr – noton – back, L-chorda-cord). This is found in all the members of the phylum at some stage of their lives. Chordates are animals having a cord are "notochord" or "Invertebrates" since they have no notochord or back bone in their body structure.

Phylum chordate is the largest to the deuterostiome phyla. It is the highest and the most important phylum comprising a vast variety of living and extinct animals including man himself most of the living chordates are the well known familiar vertebrate animals, such as the fishes, amphibians, Reptiles, birds and mammals.

Three fundamental characters of chordates:-

All the chordates possesses three outstanding unique characteristics at same stage in their life history. These three fundamental morphological include,

i) A dorsal hollow or tubular "nerve cord"

ii) A longitudinal supporting red like "notochord"

iii) A series of pharyngeal gill slits.

These three distinctive characters which set chordates apart from all other phyla.

phyla.

Phylum chordate is conveniently separated into 3 to 4 primary subdivision called subphylum: based on the characters of " Notochord" these are.

Subphylum – Hemichordate/Adelochordata

Subphylum – Ucochordata/Tunicata

Subphylum – Cephalochordata

Subphylum – Vartebrata.

Classification proposed by Berg (1940) :

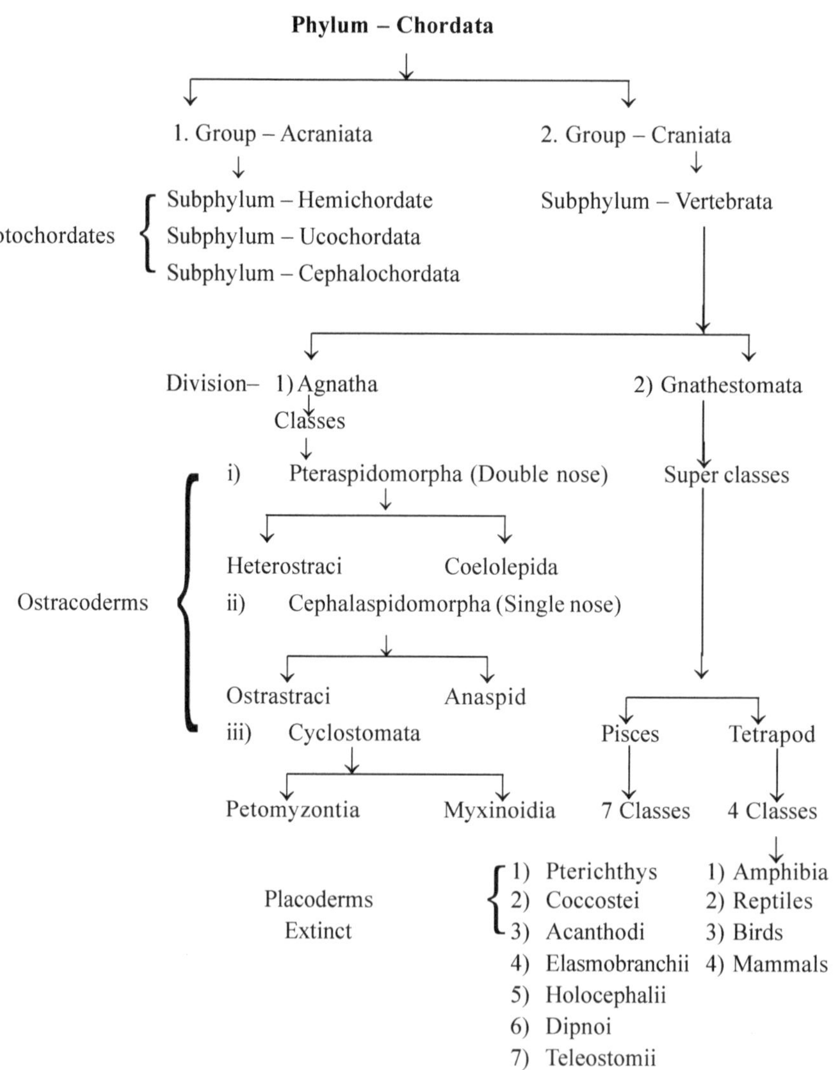

Subphyla

1. Subphyla – Hemichordata (Gr- hemi –half, chorde-cord) have long been traditionally considered to be the lowest chordates. But recent different workers consider the so-called notochord of the hemichordates not a true notochord but a stomachord, Hence the hemichordates are removed from the chordates and treated as an independent invertebrate phylum.
2. Subphyla urochordate (Gr oura – a tail, L – chorde – card) in this subphylum animal notochord is present in the tail region only. Hence it is called as uro chordata
3. Subphyla cephalachordata (Gr – kephale – head, chorde – cord) in this subphylum species notochord is present only in head region only. Hence it is called as caphalochord called as protechordats.
4. Subphyla – vertebrata (L – vertebratus – back hone) vertebral column is present in these animals hence it is called vertebrates. This subphylum is sub-divided into nine classed as ostracoderrni. reptella, aves and mammalian.

Protochordata and Euchordata

The first two subphyla under phylum chordate is urochordata and caphalochordata are all marine relatively small and without a vertebral column or back –bone. They are often referred to collectively as the non – vertebrates or invertebrate chordates or protochordates car-protos- First, chorde – card) as they are regarded to be early primitive, borderline or 1 st chordates closely allied with the ancamstrat chordat stock. At present the hemichordate are placed as an independent invertebrate phylum.

The last subphylum is, vertebrata, provided with a vertebral column is regarded to be more advanced and belongs to the subdivision Euchordata of phylum chordate, sometimes the protochordates are known as the lower chordate, while the vertebrates as the euchordates as the higher chordates.

Group – Acraniata and Craniata

The protochordate subphyla lacks a head and a cranium i.e. brain box, so that they are known as Acraniata (Gr –a-absent, kranion – head)

On the other hand the subphylum vertebrata has a distinct head and a cranium is brain box and so it is called as craniata.

Divisions – Agnatha and Gnathostomata :-

The vertebrata or craniata are furthen sublivided in several ways. One possibility with universal agreement separates them into two unequal sections. Agnatha and Gnathostomata agnatha (Gr-a-not, gnathos-jaw) lack true jaws and paired appendages. Agnathas or agnathostomatas include a small number of primitive but highly specialized Fish – like forms the extinct ostraoderms and the modern cyclostomes.

And all other vertebrate have true jaws and paired appendages and are called as Gnathostomata (Gr-gnathos-jaws, stomata-mouth). These are also called as jawed animals.

Pisces and Tetrapoda

A basic division of Gnethostermata recognized two super-classes or series which are pisces and tetrapoda. The super-class pisces (L-piscis – Fish) includes all the fishes which are strictly aquatic forms with paired fins.

The super class Tetrapoda (Gr-tetra-four, podos-foot) is formed by four – legged land vertebrates including amphibians, reptiles, birds and mammals.

(i) Pisces :- super-class or series

The super-class or series pisces are divided into seven classes. These are aquatic cold blooded animals, which respires with the help of gills and swims with the help of fins. Among these classes 1'st three classes is pterichthyes, coccosteus and Aconthodi, these are extinct classes. These extinct classes collectively known as placoderms, because these are early sawed fishes or vertebrates that appeared during the Silurian and were quite dominant for a brief period during the Devonian but all of them become extinct by the end of the Permian period

They lived during the Paleozoic era only and are sometimes referred to as the Paleozoic fishes. All of them were well protected by heavy, plates or bony armor and are therefore called as "placoderms". Remaining four classes as elasmobranch, Holocephali, Dipnoi and Teleostomi still survive today.

1) Class – Pterichthyodes
2) Class – Coccosteus **Placoderms - Extinct**
3) Class - Acanthodii
4) Class – Elasmobranchii/Cartilagenous/chordrichthyes fishes

Characters

1) This class includes a large number of marine fishes commonly called as the sharks, skates and rays.
2) These are also called as cartilaginous fishes.
3) Endoskeleton of these fishes is made up of cartilage.
4) Exoskeleton is in the form of dermal denticles or placoid scales, arises from the plate like bases consisting of bony tissue.
5) There are 5-7 pairs of gill silts of which 1st pair change into spiracles.
6) Operculum or gill cover is absent.
7) The jaw suspension is amphistylic in condition.
8) Tail is unequal i.e. Heterocercal in condition.
9) Cloaca is present.
10) Males generally possess a pair of copulatory organ i.e. claspers.
11) Air – bladder is absent.
12) Intestine possesses a spiral valve.
13) Veins enlarged at several places to form the sinuses.
14) Fertilization is internal and advanced in relation to other fishes.
15) Both, the oviparity and the viviparity are found in cartilaginous fishes.
16) Marine elasmobranches retain a high concentration of urea in blood for osmoregulation.
17) Nostrils are one pair and there are three pairs of semicircular canals.

    Ex-Scoliodon. Trygon, Raja etc.

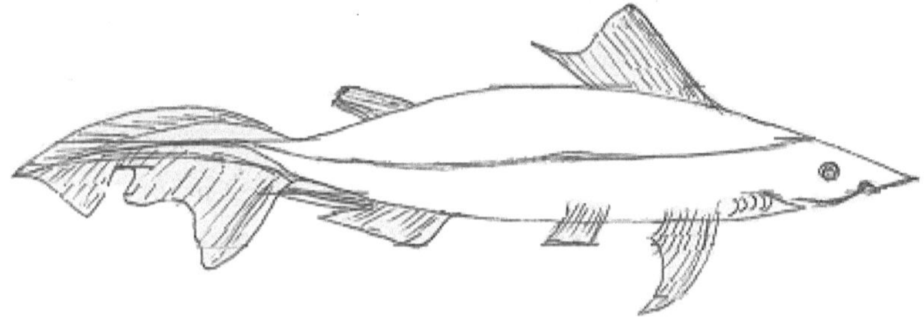

Fig 2.1 : Diagrammatic Sketch of Scolidon fish.

5) Class – Holocephalii

(Holos – entire, Kephale – Head)

Characters

1) The holocephlil are a very ancient group of highly specialized marine fishes.
2) Palato – quadrate fused to the neurocranium.
3) Four gill slits on each side and opens a single gill opening which are covered by Fleshy operculum.
4) The mouth in holo cephalian is small a ventral as compared to the mouth of Elasmobranchs.
5) Cloaca and spiracles are absent.
6) Body is naked i.e. scales are absent on body.
7) Single nasal opening is present.
8) Jaws with tooth plates.
9) The kidney of holocephalian is ophisthonephric.
10) Lateral line system with open groove.
11) Males with copulatory organs as claspers.
12) The holocephalian are oviparous, Then eggs are characteristic spindle shaped and enclosed in horney capsule
13) Fertilization is internal and the cleavages are holoblastic.

    Ex – Rat fish (chimaera), Rabbit fish and elephant fish.

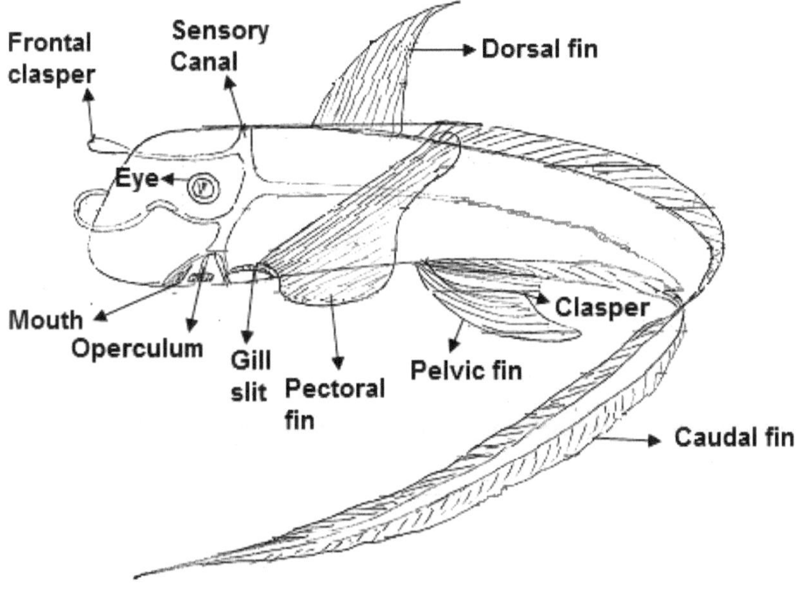

Fig. 2.2 : Diagrammatic Sketch of Chimaera – Rat fish

6) Class – Dipnoi/Lung Fishes
(Di-double, pnoi-breathing)

Characters

1) It is characterized by lobate paired fins, which overlapping the cycloid scales.
2) It is most interesting group of bony fishes and shows close affinity to the telecasts on the one hand and amphibians on the other side.
3) They arose in the mid-Devonian period and flourished during the Permian and Triassic period.
4) At present the dipnoi are represented by three genera.
5) Median fins continuous to form diphycercal tail.
6) Premaxillae and maxillae are absent.
7) Spiracles are absent.
8) Internal nares are present.
9) Air bladder single or paired lung like. Single – Neoceratodus and double or paired in lepidosiren and protopterus.

10) Body is somewhat eel like appearances.

11) Jaws pearly ossified.

12) Paired fins are supported by dermal fin eays which are joined branched and formed of a bony structure.

13) Cycloid scales are present on the body.

14) Operculum, air blader and cloaca are present.

15) The air bladder is used as accessory respiratory organ hence it is called as lung fishes.

16) Jaw suspension is autostylic.

17) In the lung fishes the heart is almost three chambered like the amphibians.

18) One pair of mesonephric kidneys is present.

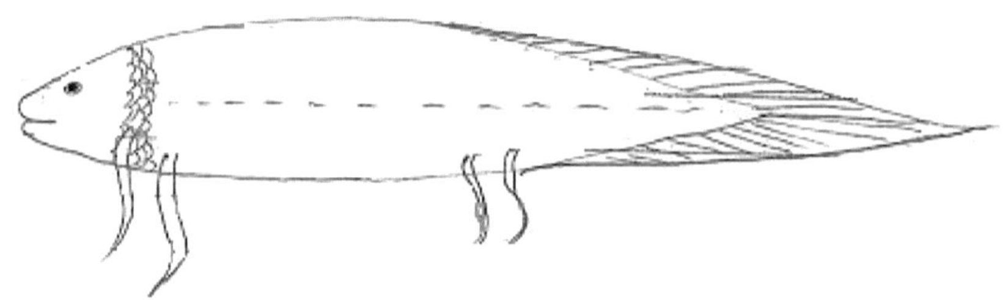

Fig. 2.3 : Diagrammatic sketch of Protopterus

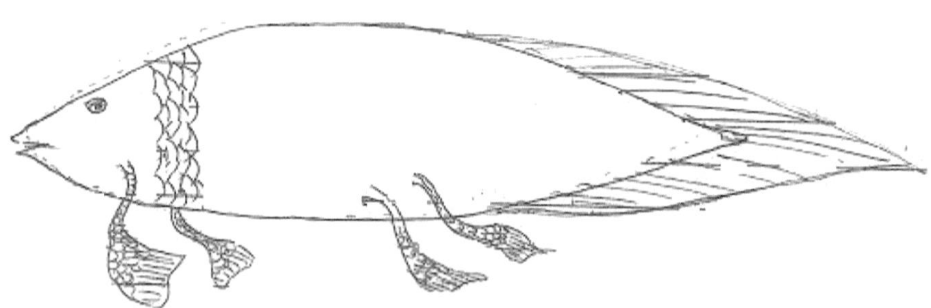

Fig. 2.4 : Diagrammatic sketch of Neoceratodus

7) Class – Teleostomii/Bony fishes/Osteicthyes/fresh water fishes

Characters:

1) These are highly advanced group of fishes.
2) Teleastomii is also called as bony fishes or osteichthyes.
3) The endoskeleton is made up in the form of bones.
4) This group includes a large number of bony fishes in which the primary upper and lower jaws are supplemented by the addition of membrane Bones, which form the secondary jaws. Hence they are called as teleostmii or perfect mouthed fishes.
5) The skin of teleostomii is usually hyostylic or amphistylic.
6) An operculum covering the gill cleft is always present and there is a single external bronchial aperture.
7) Palato – quadrate not bused with endocranium.
8) The body of these fishes is generally covered with exoskeleton and exoskeleton is in the form of rhombic or cycloid scales or ctenoid.
9) There is no cloaca these bearing separate anal and urinogenital openings.
10) Claspers or male copulatory organs are not present.
11) Air bladder is usually present.
12) Spiral valve are absent.
13) Fresh water fishes retain a high concentration of ammonia hence it is Ammonotelic animals.

3

Classification of Fishes (Berg : 1940) up to Class Level

The evolutionary classification based largely on the typological methods has been adopted a number of fish taxonomist. Foremost among them are Gunther and Cope (1871) Boutenger (1904); Jordan (1923); Regan (1929); Berg (1940) and Romer (1959). Out of the various schemes proposed by aforesaid taxonomists, there is no agreement among the ichthyologists regarding the classification of fishes. Fishes constitute more than one half of all vertebrates comprising around 40,000 species.

The classification of Berg (1940) is most widely accepted, it is based upon the conclusions of Jordan (1923) and Regan (1929). Many ichthyologists consider Bergs classification as a minor revision. Berg scrutinizing the merits of the other classification proposals and make advanced classification, which is accepted all over the world.

Classification Proposed by Berg (1940)

Berg (1940) included all the living and the fossil fishes within a series – Pisces, and subdivided the series into seven major classes. The class pterichthyes, coccoestri and aconthodi included all the primitive fishes that are collectively known as placoderms. The class Elasmobranchii consisted of cartilaginous sharks, skates, and rays, class Holocephali included a small group of aberrant fishes called as chimezoides, dipnoi, included three different genera of lung fishes and the class Teleastomii the remaining are bony fishes. The outline classification proposed by Berg (1940) is given in detail in Chapter 2. The details of each class from the 7 classes proposed by Berg are given below:

Class – I – Characters of Pterichthys

1) Pterichthys are also called as antiarchi.
2) Head and the anterior part of the body covered with bony carapace consisting of symmetrically arranged large plates.
3) Pectoral fins are puddle like with a bony covering.
4) Tail is heterocercal.
5) One or two dorsal fins are present.
6) Eyes are placed close together on the dorsal surface and there is a pineal opening in between them.
7) Nostrils are two in number on either side.
8) The head and the trunk shield were made up of closely united dermal plates.
9) Mouth was ventral and jaws are weakly developed.
10) Absence of pelvic fins and toothed jaws.
11) Dorsal fin is spinous.
12) This class is extinct from the earth planet.

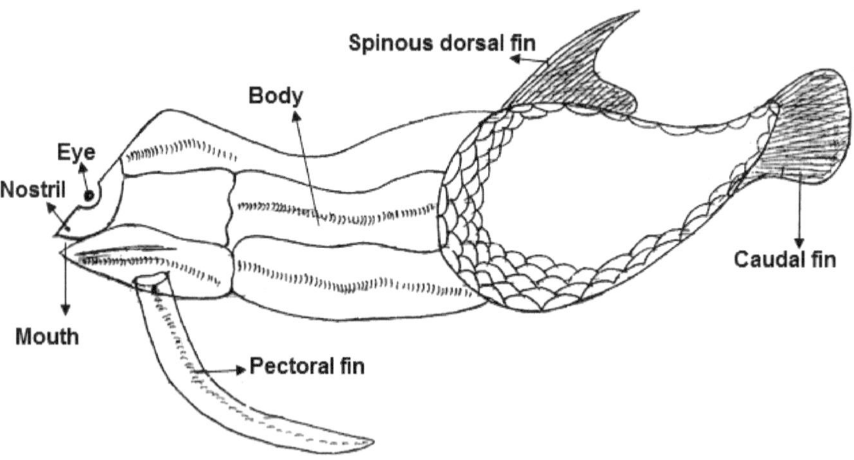

Fig 3.1 : Diagrammatic sketch of Pterichthys

Class – II Characters of Cccosteus/Coccostei/Arthrodira

1) The fossils of these armed plated fishes are abundantly found in the Silurian and the Devonian Periods or rocks.
2) Their overall appearance is strongly reminiscent (reminding) one of the modern skates and rays.
3) It is characterized by two plates is cephalic plates and thoracic plates.
4) Head and anterior part of the body is covered with a bony plates or carapace consisting of large plates. Head carapace is movably articulated with that of the body.
5) External gill aperture is present between the head and body carapace.
6) Notochord is persistent for long period
7) Neural and Haemal arches are ossified.
8) Pectoral fin is replaced by immovable spines
9) A dorsal fin is present.
10) Eyes are laterally placed with a pineal aperture in between them.
11) The tail were small whip like and heterocercal covered with small scales.
12) This class is also extinct on the earth planet.
13) These are only fossil fishes.
14) This class fishes are at called as placoderms.

    Ex- Coccosteus.

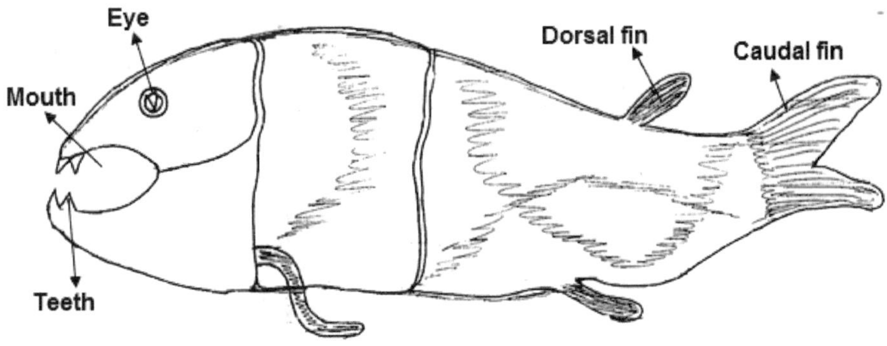

Fig 3.2 : Diagrammatic sketch of Coccosteus.

Class – III – Characters of Acanthodi

1) Gr. word – acontha – meaning spines.
2) This class is also extinct class.
3) This class fishes are also called as placoderms.
4) Presently these class fishes are found only in fossil records.
5) The acanthodians are the earliest known jawed fishes.
6) The development of jaws in these fishes is regarded as the greatest of all the advances in the vertebrate history.
7) Most important character is the presence of stout spines
8) The head was large and blunt and had a pair of long and well set eyes at its anterior ends.
9) Entire body was covered with small squarish scales i.e. ganoids scales are present.
10) The mouth was large and suited with sharp cutting teeth.
11) Endoskeleton consisting of true bones.
12) A gill slits is present between the mandibular arch and the hyoid arch.
13) Four or Five branchial arches are present.
14) Notochord is persistent stay for long time.
15) Caudal fin is heterocercal in condition.
16) One to five pairs of spines are present between the pectoral fin and ventral fin
17) An operculum as external covering is present.

    Ex-Climatius (an acanthodian).

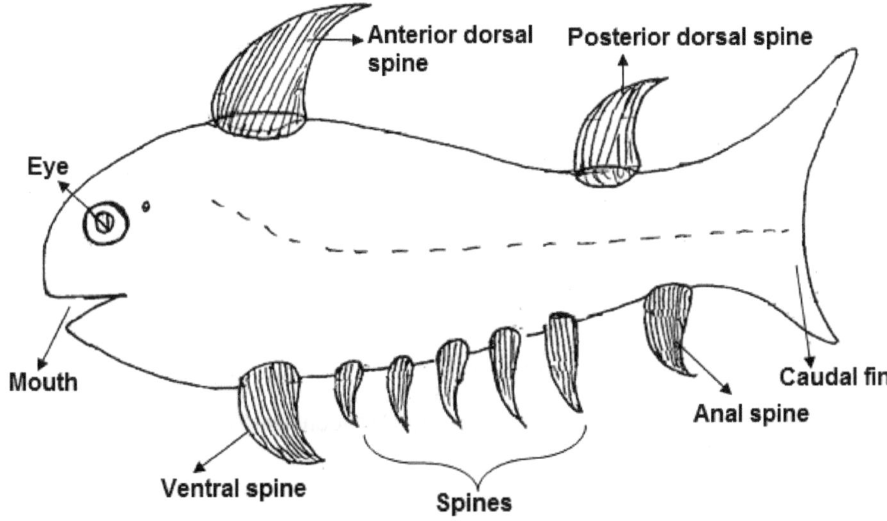

Fig 3.3 : Diagrammatic sketch of Climatius

Class – IV: Characters of Elasmobranchii OR Cartilagenous fishes Of Chondrichthys OR Marine water fishes

1) These fishes appeared in Devonian period and survived a long period of nearly 400 million years.
2) Elasmobranchi fish is very primitive type of fishes.
3) Elasmobranchi fish is also called as cartilaginous fishes.
4) These fishes are also called as chondrichthys fishes.
5) In this class species, maximum fishes survives in the marine water medium hence it is called as marine water fishes.
6) These fishes are primitive and less advanced among the fishes.
7) This class includes – large number of marine fishes commonly called as sharks, skates and rays.
8) These fishes having no bones other than these of the scales (placoid) and teeth.
9) These fishes are also called as marine water fishes as well as cartilaginous fishes as well as chondrichthyes fishes.
10) Endoskeleton of these fishes is made up of in the form of the cartilage ribs, hence it is called as cartilaginous fishes.

11) Exoskeleton is in the form of dermal denticles or placoid scales arises form the plate like bases consisting of bony tissue.

12) There are 5-7pairs of gill slits of which I[st] pair change into spiracles. The gill slits remain naked and visible because they uncovered by a fold of operculum.

13) Operculum or gill cover is absent in these fishes.

14) The jaw suspension is amphistylic or hyostylic in condition.

15) The brain of eleasmobranchi is large in relation to the size of the body.

16) The tail is unequal condition is Heterocercal tail.

17) Marine elasmobranchi retain a high concentration of urea in blood for Osmoregulation, hence these fishes are also called as ureotelic animals.

18) Veins are enlarging at several places to form the sinuses.

19) Air bladder in this class is entirely absent.

20) Cloaca i.e. opening for fecal matter, urine and genital process is present.

21) Males generally possesses a pair of male copulatory organ at the base of pelvic fins called as claspers.

22) Intestine possesses a spiral valve.

23) Fertilization is internal and advanced in relation to other fishes.

24) Both the oviparity and the viviparity is found in this class fishes.

25) Nostrils are one pair and there are three pairs of semicircular canals in inner ear.

26) Elasmobranchs are usually prediceous in nature.

27) Mouth in these fishes is on ventral side.

Example sharks, skates and rays.

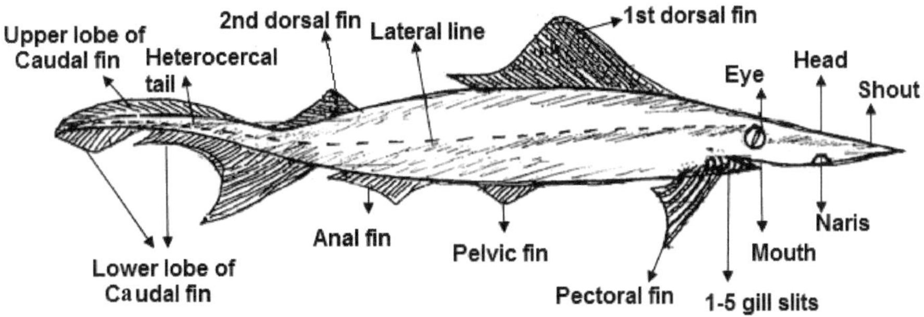

Fig 3.4 : Diagrammatic sketch of Scoliodon (Shark)

Class – V – Characters of Holocephali

(Gr. word Holos – Entire, kephale – Head)

1) The holocephali are a very small ancient group of highly specialized marine fishes.
2) These fishes have a shark – like body with a relatively large compressed head, a small mouth and large eyes.
3) Endoskeleton is cartilaginous often calcified.
4) Notochord is persistent and poorly developed vertebrae.
5) Palato – quadrate fused to the neurocranium.
6) Four gill slits on each side and opens a single gill opening which are covered by fleshy operculum.
7) The mouth in holocephalian is small and ventral as compared to the mouth of elasmobranchs.
8) Skin is smooth silvery and placoid scales occur in patches, as in chimaera, in other fishes body is naked i.e. scales are absent.
9) Spiracles are absent and eyes are large in size.
10) The urinogenital aperture and the anus are separate, and there is no cloaca.
11) Single nasal opening is present.
12) Jaws with tooth plates teeth united to form crushing plates, devoid of enamel this is an adaptation for crushing animals.
13) Jaw suspension is holougt stylie in the upper jaw is immovably united with cranium, hence the name holocephali.

14) Median and paired fins are well developed. And dorsal fin in chimaera is fairly long, pectoral fins are long and pelvic are smaller.

15) Tail is generally heterocercal but in chimaera it appears to be diphycercal and in the form of a whip like.

16) Kidney is opisthonephric corresponding fundamental pattern of fishes.

17) Lateral line system is well developed with open groove in chimaera and closed groove in other fishes.

18) Sexual dimorphism is well marked, males with copulatory organs called as claspers.

19) A peculiar frontal clasper is present on the head of the male chimaera. The function of this clasper is not known.

20) The holocephalians are oviparous, their eggs one characteristic spindle shaped and enclosed in horney capsule.

21) Fertilization is internal and the cleavages are holoblastic.

Examples – Chimaera (ratfish) rabbit fish and elephant fishes.

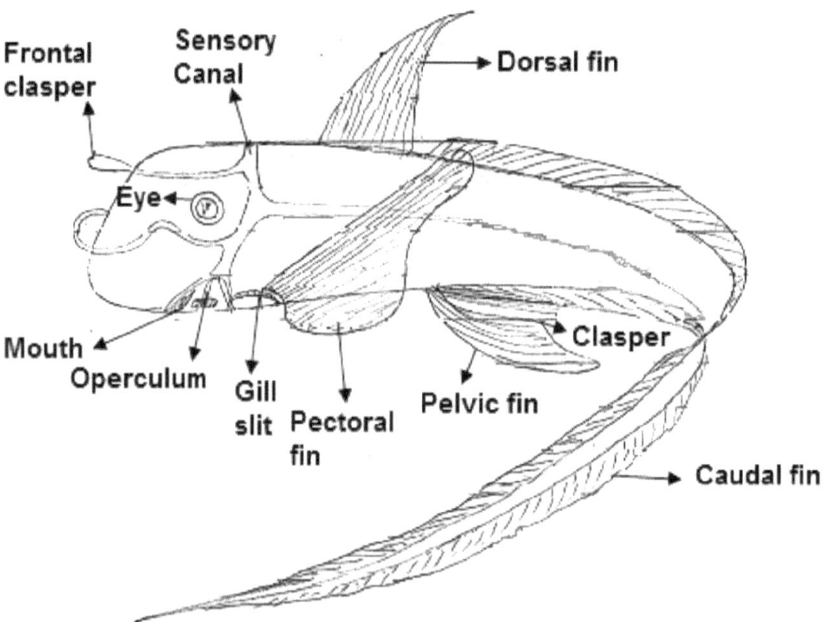

Fig 3.5 : Diagrammatic sketch of Chimaera – Ratfish

Class – VI – Characters of Dipnoi/Lung Fishes

(Gr. word -di-double : pnoi-breathing)

1) The Dipnoi are also called as the "lung fishes"
2) The dipnoi fishes are appeared in the mid- Devonian period, and flourished during the Permian and the Triassic.
3) At present, only three genera of Dipnoi exist on the earth planet as protopterus of Africa, Lepidosiren of S. America and Neoceratodus of Australia.
4) It is most interesting group of bony fishes and shows close affinity to the teleosts on the one hand and amphibians on the other hand.
5) It is characterized by the lobate paired fins which overlapping the cycloid scales but the fossil forms had cosmoid scales.
6) The median fins continuous to form diphycercal tail.
7) Jaw suspension is autostylic and the palatoquadrote is fused with cranium.
8) Notochord is persistent and unconstructed.
9) Internal nares are present.
10) Premaxillae and maxillae are absent.
11) Spiracles are also absent.
12) Body is somewhat eel like appearance.
13) Mouth lies on the ventral side of the head.
14) Dipnoi is unique in possessing both aquatic and aerial respiratory organs.
15) Air bladder single or paired lung like single in Neoceratodus and double or paired in lepidosiren and protopterus.
16) External gill opening is present and covered by operculum.
17) Cloaca i.e. all openings are combines which is present.
18) A pair of mesonephric kidneys is present.
19) Paired fins are supported by dermal fin rays which are jointed branched and formed of a bony structure.
20) In the lung fishes the heart is almost three chambered like that of amphibians.

21) The air bladder is assessor respiratory organs used hence it is called as the lung fishes.

22) Endoskeleton is largely cartilaginous.

Example – Neoceratodus Lepidosiren protopterus.

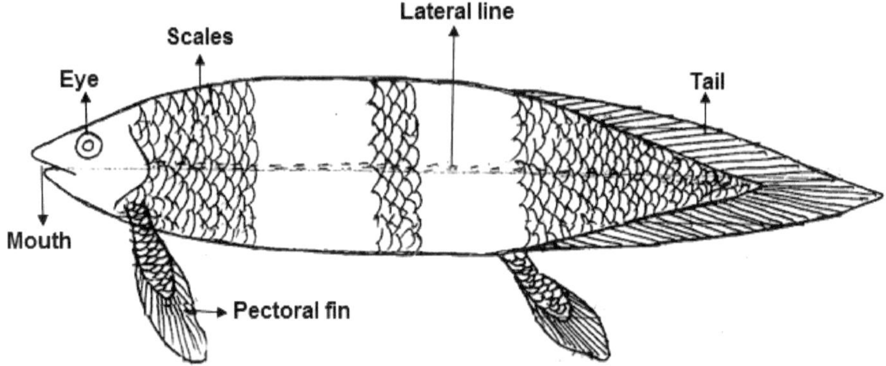

Fig 3.6 : Diagrammatic sketch of Neoceratodus

Class VII – Characters of Teleostomii/Bonyfishes/Osteichthyes/ Fresh Water fishes.

1) These are highly advanced group of fishes.
2) Teleostomii is also called as bony fishes or ostrichthyes.
3) Large number of bony fishes live in fresh water medium hence these are called as freshwater fishes.
4) The endoskeleton is made up of in the form of bones.
5) These fishes are flourished during the Devonian period, Devonian period is also called as Age of fishes.
6) In this fishes the primary upper and lower jaws are supplemented by the addition of membrane bones, which forms the secondary jaws, hence they are called as perfect mouthed fishes or teleostomi.
7) The skull of teleostomii is usually hyostylic or amphistylic.
8) The body of these fishes is generally covered with an exoskeleton and Exoskeleton is in the form of rhombic of cycloid or ctenoid scales
9) Single external branchial aperture is present called as operculum.
10) Palato – quadrate is not fused with endocranium.

11) There is no cloaca, there being separate anal and urinogenital apertures.

12) Claspers or male opulatory organs are absent in these fishes.

13) Air – bladder or swim bladder is usually present.

14) Spiracles are absent in this class.

15) Gill filaments project beyond the interbranchial septum which is greatly reduced.

16) Spiral valves are absent in the intestine of these fishes.

17) Fresh water fishes retain a high concentration of ammonia in the urine, hence, it is called ammonotelic animals.

18) Both oviparity and ovivparity is seen in this group.

19) Fertilization is external outside the female body in water medium.

20) Kidneys in teleostomii are mesonephric.

21) Tail of teleostomii is both homocercal as well as heterocercal.

22) The brains one small in relation to the size of the body.

23) There are 4-5 pairs of gill slits are present.

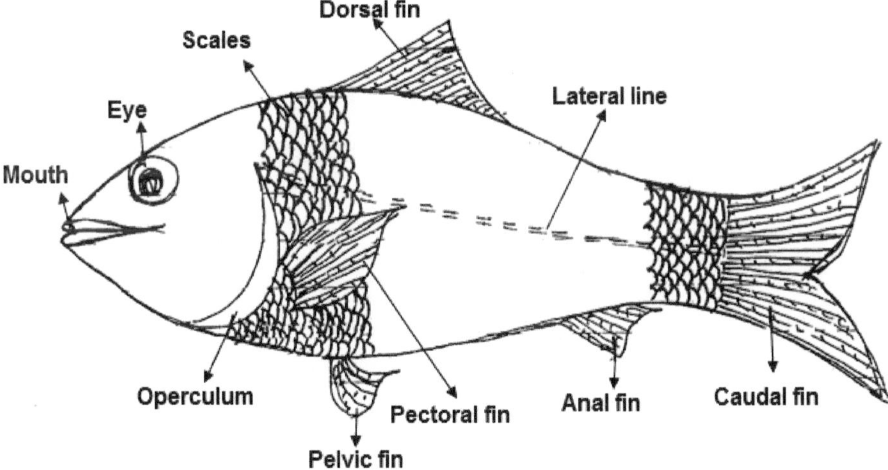

Fig 3.7 : Diagrammatic sketch of Bony fish

4

External Characters of Teleosts

This group includes a large number of bony fishes in which the primary upper and lower jaws are supplemented by the addition of membrane bones which forms secondary jaws. Hence they are called as teleostomi or perfect mouthed fishes.

All major groups of fresh water fishes lived successfully and flourished during the Devonian period which is also called as the "Age of fishes ". Teleostomii are characterized by a unique mechanism of opening the mouth by lowering the mandible through hyoid apparatus, therefore the skull is hyostylic or the amphistylic. Presence of an ossified operculum, an interhyal bone and the branchiostegal rays are the diagnostic features of this group.

Characters of Teleosts

The external characters of teleosts are studying with following various parts as the body of teleost fish is divisible into three part as Head, trunk and tail.

1. Head

Body of some fishes are elongate; sub-cylindrical, cylindrical body flat end, of dorsoventrally flattened or compressed in some fishes body is eel like appearance .In some teleosts head is long and compressed or in some fishes it is depressed. In same fishes head is short where as in few fishes head is very long. In same teleosts head width is same to its length and in few it is less. In few fishes in front of head a snout is produced the snout is pointed or blunt.

In the head region of fish various parts are present as in few fishes lower jaw is longer than upper jaw. In some fishes upper jaw is longer as compared to lower jaw on head of few fishes scales one present where as in some fishes head is covered by thick skin without scales. In various fishes on both jaws teeth are present which are the vallifom teeth.

In few fishes teeth are present only on one jaw either the upper jaw or lower jaw, while in other fishes cardiform teeth are present. In some fishes mouth is terminal or sub terminal and crescentric opening. In some teleosts mouth is inferior or lower side and mouth opening is small while in some fishes large. In few teleosts barbles are present which are one pair or in two pairs as maxillary barbles and the mandibular barbles.

In maximum teleosts barbles are absent. In few when present it is very short, thick barbles. The eyes in only some fishes are not covered with eyelids .In some teleosts nostrils are present just in front of barbles which are one or two.

In teleosts the typical opercula opening or gill opening or the bronchial opening is present just behind to the eye. External bronchial or gill aperture is the main identification characters among teleastomic and elasmobranchii. The operculum or gill opening protected the gill which is present inside it, gill opening in some teleosts is wide and in few it is small. In some teleosts as catla the gill opening is very wide extend up to cleft of mouth.

2. Trunk

Trunk is the second part of fish body. In some telcosts trunk is longer part where as in some fishes it is very small part. In some teleosts trunk is flattened or dorsoventrally compressed. Trunk forms maximum muscle bundles or myomeres more myomere bundles more will be the fish and vice- versa. The trunk extends from the posterior border of the operculum or just behind the loose gill slits to the anal aperture or opening. In some fishes trunk is thinner where as in few teleosts it is thickest part of the body. The trunks of some teleosts are covered by thick skin and scales are absent on skin and it is naked, where as in few teleosts total trunk is covered by skin which possesses varies types of scales. The scales are of either cycloid or ctenoid scales on the body. In some teleosts the scales are very minute and small where as in few the scales are very large. Scales are larger on dorsal side where as it is smaller on ventral surface of fish body.

The trunk bears five fins and three opening or apertures. The fins are thin flat outgrowths of the skin with muscles and supporting rods called as the fin rays. In some fishes fin rays are thin delicate where as in other teleosts it is hard and tough. The fins are of two types i.e. median or unpaired fin and lateral or paired fins. When there is a single or median fin present on the fish body it is called as median or unpaired or single fin such as dorsal fin, caudal fin and anal fin. Various types of fins are described below:

a) Dorsal fin- The outgrowth of the skin with muscles arises on dorsal side of the fish body hence it is called as dorsal fin. In some teleosts two dorsal fins are present. But in maximum teleosts only one single dorsal fin is present. In few teleosts in the dorsal fin spines are also present then it is called as spinous dorsal fin. In some fish spines are absent only simple fin is present. When two dorsal fins present then those two dorsal fins are called as first dorsal and second dorsal fin.

In some teleosts instead of second dorsal fin the fin is present but it is flexible and without fin rays. If there are thin muscles are present that fin is called as adipose fin. Such adipose fin is present in few members of telcosts.

b) Cadual fin- The fin arise or develop on posterior most part of tail of the fish or caudal end that fin is called as caudal fin. It is also unpaired or single or median fin .The caudal fin is of two types again as homicercal caudal fin and heterocercal caudal fin. Homocercal when both lobes of caudal fin is same or similar it is called Homocercal. When one lobe of caudal fin either upper or lower is longer or larger, well developed or another is smaller or shorter then it is called as heterocercal caudal fin. In caudal fin spines are absent.

c) Paired fins- The Paired or lateral fins .The fins when develop or arise on both sides of the fish body that fin is called as paired fin or lateral fin. Paired are of two types i.e. (i) pectoral fin and (ii) pelvic fin or ventral fin.

(i) Pectoral fin- The fin originating just behind to the operculum on both sides that fin is called as pectoral fin. In few teleosts pectoral also possesses the strong spine which is also serrated or non-serrated. Pectoral fin in some teleosts are longer than dorsal and anal fin.

(ii) Pelvic fin – The fin placed or originated on ventral side just on abdomen region and which is a smaller fin is called as pelvic fin or ventral fin. It is also a spinous or nonspinous pelvic fin.

On the trunk region three apertures or openings are present. The anal opening or aperture is a median simple rounded opening, just behind the posterior end of the base of the pelvic fins.

Just in front of the anal opening or aperture is present it is called as genital opening or aperture and just in front of the genital aperture the opening is present it is called as urinary opening or aperture.

3. Tail

The last part of the body of fish is called tail. In some teleost tail is very short where as in other fishes it is very long. Tail can be compressed in some varieties and can be flattened or laterally depressed. Tail of teleosts forms two median fins as caudal fin and anal fin.

a) Caudal fin:- The fin arise or develop on posterior most part of the body of fish is called as tail or caudal end that fin is called as caudal fin .It is also median or unpaired fins. The caudal fins are of two types as homocercal and heterocercal caudal fin. When in some teleosts both lobes of caudal fin are same or similar it is called as homocercal caudal fin. When one of the lobe of caudal fin is larger and well developed and one lobe is smaller or shorter than it is called as heterocercal caudal fin. In caudal fin spines are absent.

b) Anal fin :- In front of the anal opening or aperture the fin starts, that fin is called as anal fin. In few teleosts anal fin is long where as in few teleosts it is very short. The spines are also present in few teleosts and in some fishes it is absent. In tail region muscle bundles are thin and flat.

A clear and prominent lateral line runs on either side of the body, from just behind the opercula opening to the base of the caudal fin more towards the dorsal line is complete and it is interrupted line.

All these above are the external character of the teleost fish.

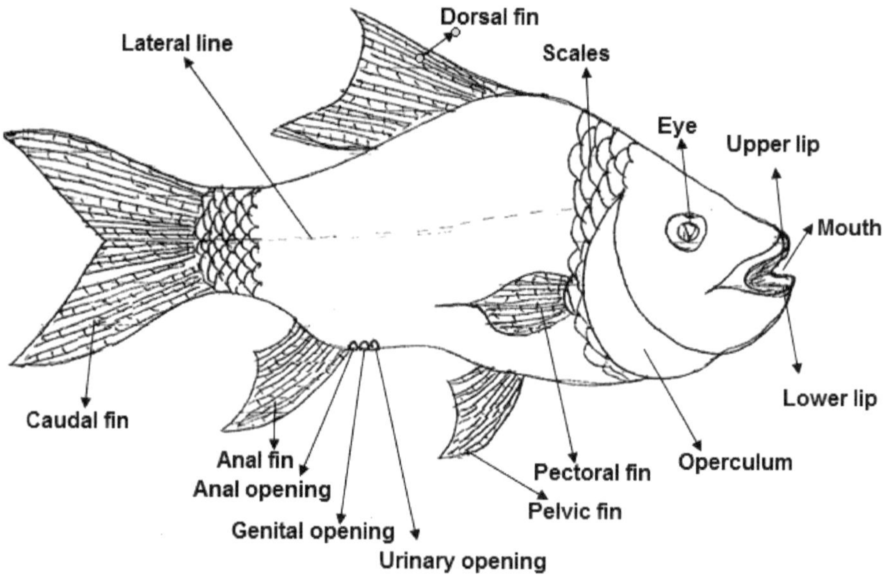

Fig. 4.1 : Diagrammatic sketch of Teleost fish showing external characteristics.

5

Body Forms and Locomotion in Fishes

The animals, which are aquatic cold blooded respires with the help of gills and swim with the help of fins such animals are called as fishes. Fishes have various body forms according to the environment and in changing the life of fish the body form of fish are found. The fish swim in water with the help of fins. The fins are present on all body of the fish. In swimming the body form is also important. Following are the various body forms found in fishes.

Body Forms

Following are some of the body forms in fishes:

1. **Anguilliform body from:-** When in fishes the body form is the anguilliform shape and size, like that of eel or anguilla type, then such body form is called as angilliform body form.
2. **Depressiform body form :-** When in the fishes body form is flattened from above downwards, means the body form is depressed dorso-ventrally it is called as depressiform body forms like that of angler fish, ray fish is also called dorso-ventrally depressed.
3. **Disciform body form :-** When in fishes the body form is flattened form sides or the fish is laterally compressed on both sides that type is called as disciform body form of laterally compressed body form like that of flat fishes.
4. **Globiform body form:-** When in fishes the body form is rounded or spherical in shape and size that body form is called as globiform body form or it also called as cylindrical body form. Such as thc sunfish or porcupine fish like the body form, name itself indicating the body is circular

globe like forms.

5. **Aculeiform body form:-** When in fishes the body form is needle shaped. Both ends are pointed such body form is called aculeiform body form which is found in pipe fish.

6. **Taeniform body form: -** When in fishes body is ribbon like or very elongated long such type of body form is called taeniform body form which is found in ribbon fish.

7. **Sagittiform body form:-** When in fishes the body form is dast like i.e. a small pointed missile used as a weapon. It's body form is like arrow which is pointed and move suddenly that body form is called as sagittiform body form. It is found in pike fish.

All above are the various body forms in fishes. Instead of above forms another few body forms also found are explained below:

a) **Subcylindrical body form** – When in fishes body form is not completely cylindrical, it is similar to cylindrical it is called as subcylindrical body forms.

b) **Cylindrical body form** – When in fishes body form is completely cylindrical or cylinder like that is called cylindrical body form.

c) **Spindle shaped body form** – When in fishes body is spindle shaped appearance i.e. thicker in front than behind that is called as form is completely cylindrical or cylinder like that is called cylindrical body form. Spindle shaped body form.

d) **Fusiform body form** – When in fishes the body is fusiform it is shaped somewhat like cigar, circular or elliptical in cross section and thicker in front than behind it is resembling a double wedge.

e) **Unique body form** – In some fishes the form of the body is unique i.e. like that of sea horses, chimera etc. These body forms are of unique structures and shapes as compared to other fishes.

Locomotion in Fishes

Fishes are perfectly adapted for swimming or locomotion in water. Locomotion of fishes provides a number of interesting information to the ichthyologists. The body of fish is variable and variations occur above in body forms of fishes. The knowledge of the different methods of locomotion is yet not fully understand because fishes somewhere else other than their natural inhabitation tend to behave in a manner same where different from the normal, but now with the help of cinematography fishes swim by the following methods:

1. By alternate contraction and relaxation of the muscles of the body called the myomeres. During swimming the fish oscillates from side to side showing lateral undulations.
2. By various movements of the fins.
3. By sudden expulsion action of jets of water expelled from the gill opening during the process of respiration.

Locomotion only by means of fin movements takes place when slow progress is desired but for rapid swimming, body movement is most important the paired fins serve for balancing the body.

The tail and the caudal fin are the chief locomotary organs of fish and are used for rapid swimming during which tail is lashed from side to side by alternate contraction and relaxation of the myomeres on the two sides of the vertebral column.

The locomotion of fishes provides the biologist and physicist with a number of interesting problems and has also attracted their attention.

1. **By alternate contraction and relaxation of the muscles of the body :-** Regarding this type of locomotion " The actual forward thrust is effected by the pressure of the fishes body against the surrounding water aided to a certain extent by the tail fin stroke due to the action of the muscles myomeres." The movement is initiated by the contraction of the first few myomeres on one side. The anterior part of the body is thrown into a curve and this curve is passed backwards in a series of waves by the alternate contraction and expansions of the serial muscle segments.
2. **Locomotion due to the movement of fins:-** Turning to the second of primary methods of locomotion it may be noted that all fins are very characteristic part of fishes. Fins are important in fish locomotion and without them the fish may turn sideways or rolling down of body at least at a faster speed. A great number of fishes are able to move about or make specialized movements for the swimming in fishes with a more or less round cross section like sharks, in laterally compressed fishes the caudal peduncle is of greatest importance. All of the various fins are used as primary organs of propulsion.
3. **Movement caused by the action of jets of water expelled from gill opening during respiration:-** This method may play same part in during the body forward. This method is probably to achieve high speed and

assists the muscular activities of the fish body when the fish strength is between they would clearly produce the best effect. A power full jet is usually expelled when a fish commences any swimming movements.

All above are these three important locomotary movements occur in fishes.

Types of Locomotion

There are main three types of locomotion in fishes as 1. Anguilliform type or eellike locomotion. 2. Ostraciform type of locomotion and 3. Carangiform type of locomotion.

1. **Anguilliform locomotion :-** The locomotion is serpentine in nature it is brought about by sequential, alternate contraction of the myotomes on each side of the body that locomotion is called as Anguilliform type of locomotion and it is found in eels or Anguilla fish. The movement is initiated by the contraction of fish few myomere on one side, that the anterior part of the body is thrown into a curve, which passed backwards in a series of waves by alternate contraction and relaxation of the muscle segment on each side of the body. The movement is of serpentine nature and the fish looks like a crawling snake. In this case the forward thrust is obtained entirely by the pressure of the body against the water and the caudal fin has very much reduced or absent because caudal fin is of little functions, in this type of locomotion.

2. **Ostraciform locomotion :-** In this case the head and the body are enclosed in a rigid bony case or hard bony case and the tail with a fan like caudle fin projects behind slow movement takes place by the dorsal and anal fins while the lashing movement of the tail cause rapid swimming, undulating movements are not possible that locomotion is called as ostraciform locomotion. It is a wig-way motion, seen especially in the sculling action of the tail. It is induced by simple alternate contraction of all the muscle segments on one side of the body and then on the other. The alternating contractions cause the tail to switch back and forth like a paddle behind the relatively rigid trunk of the fish. The body moves in a series of short cross arcs in a water as the fish progresses forward it is found in trunk fish.

3. **Carangiform locomotion:-** Carangifarm locomotion is the most common type of locomotion. The fish shows undulating movements of the body and the actual forward through is produced by the pressure of the tail against water that locomotion is called as carangiform locomotion. The fish drives itself forward by side by side sweeps of the tail region. It is

actually an intermediate type between anguilliform locomotion and astraciform locomotion. Such locomotion is brought about by alternate contraction of myotomes on fish one side of the body and then the other starting behind the head. It throws the body of the fish into short curves, alternating from one side to other. Thus the head end function as a fulcrum for the tail working as a flexible lever.

This carangiform locomotion occurs in various fishes.

Examples – Catla, Mrigala, Labeo, Catfishes etc.

All these three types of locomotion are found in fishes.

Body Forms and Locomotion in Fishes

Some species of fish seldom fish the body for swimming and move forward by undulating movements of the medium fins usually complete waves are seen along the fins this type of locomotion is called as the Balistiform locomotion.

Certain species as catfish parrotfish do not oscillate the body or median fins. These fin use pectoral fins for locomotion and this is called as the labriform swimming.

Special Modes of Locomotion

Besides swimming in the water, some fish species display other methods of locomotion; these are called as special modes of locomotion. Following are the special modes of locomotion generally assist the fishes in their usual way of locomotion.

1. **Jumping**—Some fishes jump to escape from enemy, for food, or for purely joy sake. The fish swimming rapidly up downs through the surface of water into the air, giving sharp flick with its tail as it leaves the water medium. All the active propulsion is provided by the muscular actions of the body while in the water.

   Ex. Mullets, marine gear etc.

2. **Flying:-** Fish jumps into the air by powering force of the tail expands its large pectoral fins and using them after the fashion of parachutes, floats of guide in air for several meters using enlarged pectoral fins stretched and rigid like an aeroplane. Both the pectorals as well as pelvic are enlarged for gliding. This type of special locomotion is called as flying.

   Ex. – Exocoetus, Cypselurus, Dachylopterus.

3. **Walking: -** Some species use their paired fins to slowly crawl on the

bottom of the water body. In these fishes the lower portion of their sizeable pectoral fins, the pectoral fins are well developed and muscular and fingerlike rays which they use for walking as an insect uses its lags on the bottom. Ex. – Cephalocanthus, Antemmarius.

4. **Crawling:-** Some fishes habitually uses the pectoral fins for crawling the pectoral fin was developed in leg like fashion which is used in crawling on the bottom of water it is called crawling. The crawling is found in angler fishes ex.- Angler fish

5. **Skipping:-** Some fishes uses its pectoral fins which are bent at an angle like elbow joint for hopping over sandy flats left bare by the repeating tides. The fishes which jump on the land with the help of pectoral fins these are called mudskipper fishes which walking on mud or ozes.

   Ex. :- Periophthalmus, Boleiophthalmus.

6. **Climbing:-** Some fishes uses its pectoral fins and opercular spins for climb up on the trees or ascending over trees these are called as climbing like perch- ex. :- Anabus

7. **Tetrapod like walking:-** Lung fishes move slowly and deliberately like mud puppy, raise themselves on their peculiar pectoral appendages and walking like that tetrapod. Ex.:- Neoceratodus, Protopterus.

8. **Burrowing:-** Some fishes burrow into mud or oze by their snouts forcibly resting into the mud or rock holes or stone holes, unless a best grips is achieved . These fishes burrowing in mud, rocks or bottom of water using snake like undulatory movements of the body. They migrate from one water body to another water body they escape during the water body using their pectoral fins and spines for this purpose. Ex.:- Eel. Channa sp, Clarius etc.

9. Some fishes have adaptive characters such as sucker fish these fishes have suckers on the dorsal surface of the head with the help of this suckers they attached to some large fast swimming fish or a boat and travels a long distances without using its own energy these fishes are called as the sucker fish. These fishes travel from one place to other places. Ex.:- Sucker fish. Above all the various types of special modes of locomotion in fishes are also found.

In structure placoid scale resembles a tooth. Each scale has consists of a basal

6

Study of Scales in Fishes

In fishes the outermost covering of the body is called as exoskeleton. The exoskeleton is covered by the covering called as skin. In most of the fishes the skin is covered with an exoskeleton in the form of scales, but a few fishes are naked having no scales on the body and body is covered only by thick skin as like the catfishes. While some other fishes like the polyodan and the Acipenser exhibit an intermediate condition, having scales or scale like plates only at same localized regions of the body. The scales of fishes are products of the activity of the skin. The scales of fishes have become modified in various ways mainly to provide protection.

The scales are useful for the classification and identification of fishes (in fish Taxonomy it is used) Fish scales are self for the determination of age of the fish. Scales are also useful for the estimation of the number of the spawning.

Scale provide important information about extinct fishes and useful in identifying food habits of piscivorous animals. Scales are important exoskeleton in fishes According to the mode of origin there are two types of scales i.e. 1) Placoid and 2) Non- placoid scales.

(I) Placoid Scales

The scale which are formed or developed due to the secretary activity of both epidermis and dermis that scale is called as placcoid scale. It is characteristically found in the elasmobranchii fishes.

Placid scales are also called as the dermal dentils, which are found in elasmobranchii fishes. These scales are very similar to the dentils of ostracoderms. They have definite aid down by osteocyte cells and the pulp cavity, ramified in a manner similar to the teeth of vertebrates.

In structure placoid scale resembles a tooth. Each scale has consists of a basal plate and a spine projecting out giving a rough surface to the skin, spine has trident structure. The basal plate is formed of a substance resembling the cement of teeth, secreted by the dermis. It is a bony plate which indicates the relic of ancestral bony armor. It is anchored into the dermis by the connective tissue fibers.

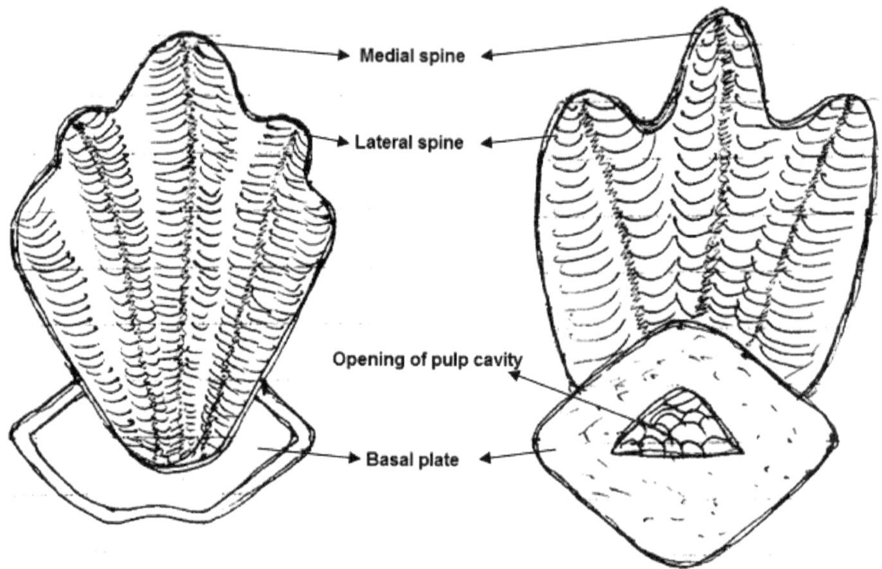

Fig. 6.1 : Diagrammatic sketch of Placoid Scale.

The spine has trident structure; the spine develops from the malpighian layer of the epidermis. The spines are curved and directed posterior thereby minimizing water friction. The outer most covering of the spine is hard transparent enamel like substance or material called as the vitrodentine.

The inner layer is the dentine which encloses pulp cavity from which several branching dentinal tubules radiate in different directions. The basal plate has an aperture in centre of the plate, which provided entrance to the blood vessels and nerves from the dermis. The pulp contains many odontoblasts means dentine forming cells fine canaliculi arise from the pulp cavity and reach the dentine.

The placoid scale does not overlap each other and are closely set in the skin. The placoid scales are partly dermal and partly epidermal in origin. Vitrodentine is epidermal and dentine is dermal in origin. Hence this scale is resembles with vertebrate teeth in basic structure. Hence placoid scale and vertebrate teeth are regarded to be homologous structures.

Ex- Elasmobranches i.e. Shark

II) Non- Placoid Scales

The scales which are formed or developed due to the secretary activity of only from the dermis that scale is called as Non-placoid scales.

Non-placoid scales are of following various types and found in teleostean fishes which are as cosmoid, ganoid , cycloid and ctenoid scales as described below:

1) Cosmoid scales

Greek word- cosmos- ornament + oide - like

The Cosmoide scales are regarded as direct descendents of bony plates of the primitive ostracoderms. Cosmoide scales are regarded as the precursor of the ganoid placoid and the bony scales of the modern Teleost. The scale is composed of four distinct layers.

The external layer of the scale is thin and enamel like and is called as the vitrodentine. The middle layer is made up of hard, non-cellular, dentine like material called as the cosmine layer which contains a large number of branching tubules and chambers. Below the cosmine layer a layer of spongy bone with numerous inter connecting canals, which containing numerous blood spaces called spongy bone layer. The innermost layer is made up of a vascular, bony substance called isopodan. It is a dense bone layer which is made of hard dentine and a thin but hard layer of enamel.

The cosmoid scales were built on a similar plan but different only in that the dentine layer way present only in the exposed part of the scale.

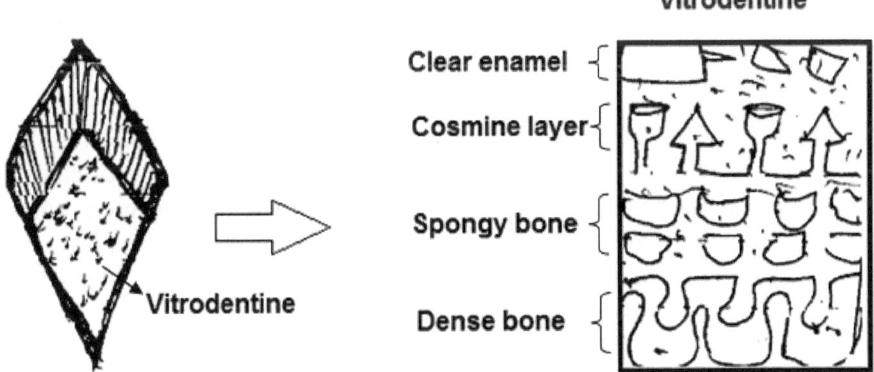

Fig. 6.2 : Cosmoid scale

These scales were abundantly found in primitive members of crossopterygil and the dipnoi. The only living members with cosmoid scale are represented by the genus Latimeria.

(2) Ganoid scale

(Greek word- Ganoi – brightness, oide – like)

Ganiod scales are heavy scales of various forms and shapes. Ganiod scales are diamond- shaped scales generally. These are thick and somewhat rhomboidal scales fit closely together in diagonal rows to cover the entire body.

These scales are heavy and have another layer of hard inorganic, enamel like material called by ganoine.

The middle layer is formed of cosmine, which containing numerous branching tubules, this layer is reduced in this scales. The innermost layer is thickest and is made up of lamellar bone called as isopedine.

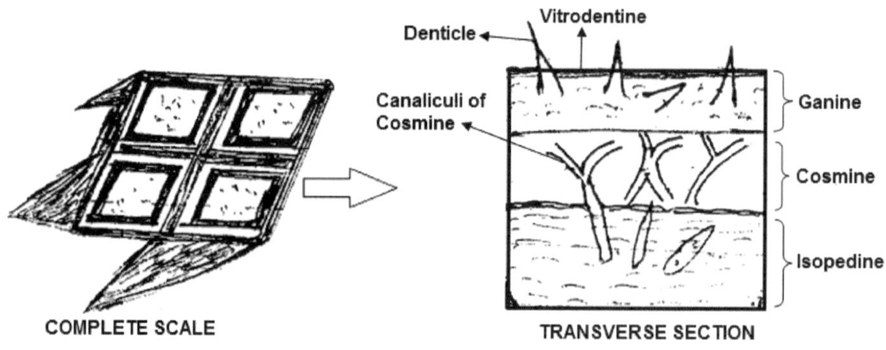

Fig. 6.3 : Complete and T.S. of Ganoid scale

These scales grow by the addition of new layer to lower as well as upper surface. These scales are usually rhomboidal in shape and articulate by peg and socket joints.

These are characteristics of the primitive actinopterygian fishes, such as Acipenser, Amia, Leipdosteus, Garpikes etc. In Lepidosteus these scales are hard rhombic-plate-like, fitting edge to edge. The middle cosmine layer is absent and it is thinner. In Acipenser these scales are in the form of large, bony scutes arranged in five longitudinal rows. In Amia, the scales have lost the genuine layer and are thin looking the like the cycloid scales typically.

(3) Cycloid scale

Cycloid scales are also known as the bony ridge scales. These are present in majority of the teleostean fishes cycloid scales are closely characteristics of soft-rayed teleports and modern Lobe-finned fishes, cycloid scales are thin, flexible and transparent, roughly rounded in shape Due to the absence of fishes layer and middle layer of cells the scale is thin and flexible with transparent structures . The free margin of the scale is rounded and entire scale is rounded and entire scale is not toothed. The scales have characteristics ridges alliterating with grooves. The ridges of the circuli are generally in the form of concentric rings.

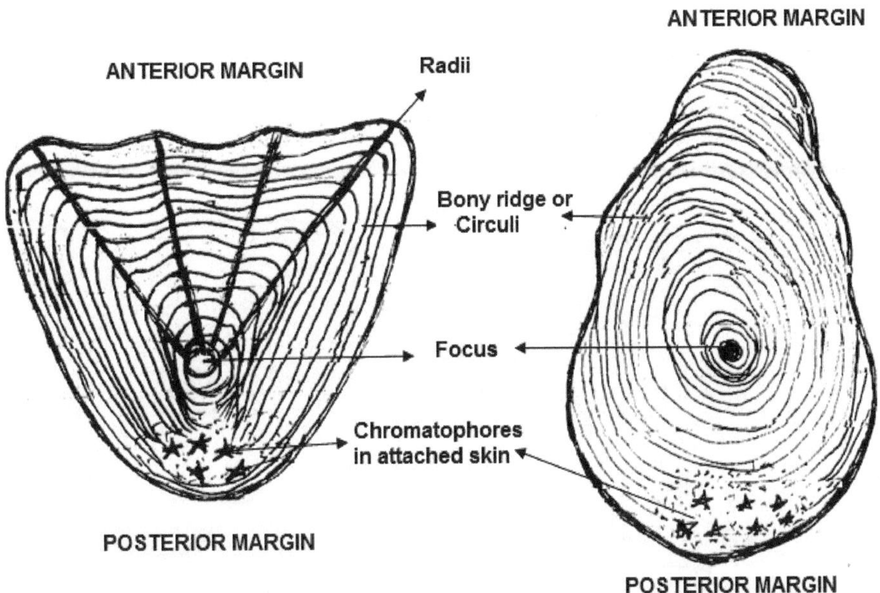

Fig. 6.4 : Cycloid Scale

The central part of the scale is thicker and is called as the focus; the scale is thinner towards margin. The central part i.e. focus or nucleus is the first part to develop. In some fish focus may be away from the centre. Focus is developing from the dermis part. The ridges or the circuli are a series of concentric bony ridges around the focus also called lines of growth or growth rings. Radii or the grooves are radiate from focus towards the margin of scales alternately. In several species oblique grooves or radii running from the focus to the margin of the scale are present.

The cycloid scales project diagonally in an imprecating pattern, forming protective covering over the body. The anterior part of the scale remains embedded in the skin, while posterior part is visible to the eye *in situ* condition. The circuli or ridges are less distinctly seen in the posterior part of the scale. In posterior part or lower part of scale chromatophores are present. The chromatophores are the colour pigments.

Around the focus circuli are develop which are more in number. Depending upon the number of circuli develop the age of fish can be studied.

The focus is first developing from dermis then the ridges are gradually added to it as the fish grows. The epidermis has no role in the scale formation. Example-carps as lebeo, catla, c. mrigal etc.

(4) Ctenoid scale

Ctenoid scale is also called as the bony ridge scale. Ctenoid scales are present in most of the spiny - rayed teleost fishes. Cteniod scales are also thin and circular like cycloid scale. The ctenoid scale differs from the cycloid in having comb like teeth or ctenior spines or more or less serrated free edge in its posterior margin of scale.

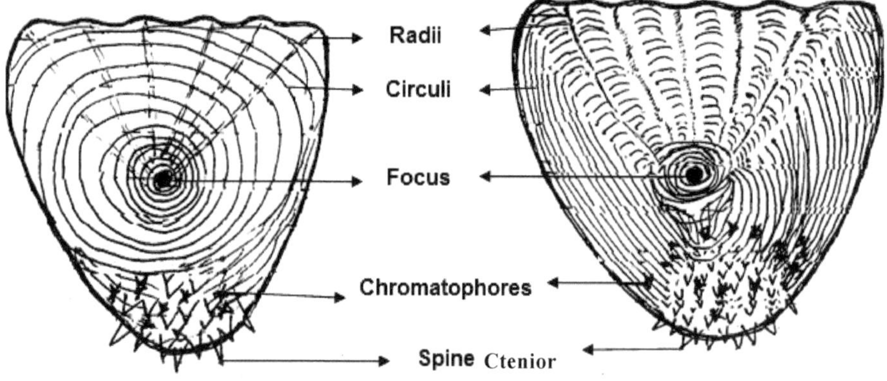

Fig. 6.5 : Ctenoid scale

Groves and ridges are alternately present. These arrange obliquely in manner that the posterior end of one scale overlaps the anterior edge of the scale present behind. These scales are composed of two lays only these are outer isopedine layer and an inner dense fibrous layer. The basal layer forms the bulk of the scale remains a cellular outside this lies a homogenous surface layer of a cellular calcified material called bony ridge layer.

The ancestral ganoine is absent and in its place a thin surface derived from enamel organ bears numerous spines, from which the scales forms its name ctenoid (Greek word centos-comb, oide - like)

The bony ridge layer scales is scalloped and bears numerous concentric ring like circuli. The circuli arrange concentrically around a central circular zone called as focus. They show well up on the embedded part of the scale and help in determining the exact age of the fish. They are fine and paced, close together when the growth of fish is slow. During rapid growth the circuli becomes heavy and fall apart. The zones of crowded and fine circulii are known as annuli. Ctenoid scales develop in the form of a dermal papilla formed by the multiplication of dermal cells of dermis, The epidermis does not take part in the formation of ctenoid scales.

Example of ctenoid scale- Anabus.

7

Electric Organs in Fishes

- Structure and working of electric organs in- Electric eel, electric catfish, electric ray.
- Mechanism of electric discharge formation and control of electric organs.

Fishes possess a number of adaptive structures which have been evolved to meet special requirements faced by them. The electric organs are highly specialized structures that enable their possessors to produce store of discharge to electric current. Fishes are unique in the animal kingdom in being capable of generating an electric current.

About 250 species of both, the cartilaginous and bony fishes of fresh and marine waters are known to possess the electric organs. Most of these restrict to the tropical fresh water of Africa and South America. Many species of rays, skates, and the eels are known to possess small electric organs that give a weak discharge. These fishes can produce electric field outside the body. Electric fishes have been divided into two types-strongly electric and weakly electric, depending on the strength of the current.

The electric organs are used for capturing prey as organs of defense, and they are also used as direction finders.

The following are the most important electric fishes with examples:

Elasmobranchii	—
Electric rays	Torpedo and narcine
Skates	Raja
Teleostomii	—
Mormyridae	Gymnarchus and memory ic
Gymnotidae	Malapterurus (Electric eel)
Siluridae	Mamapterurus (Electric catfish)
Uranoscopidae	Astroshoppus stargazer or stargazer

Structure of Electric Organs of Electric eel (Electrophorus Electricus)

This is commonly called as the electric eel and grows to length of about 8 feet. The electric organs are present on each side of this fish one of which is much larger than the other which are as below.

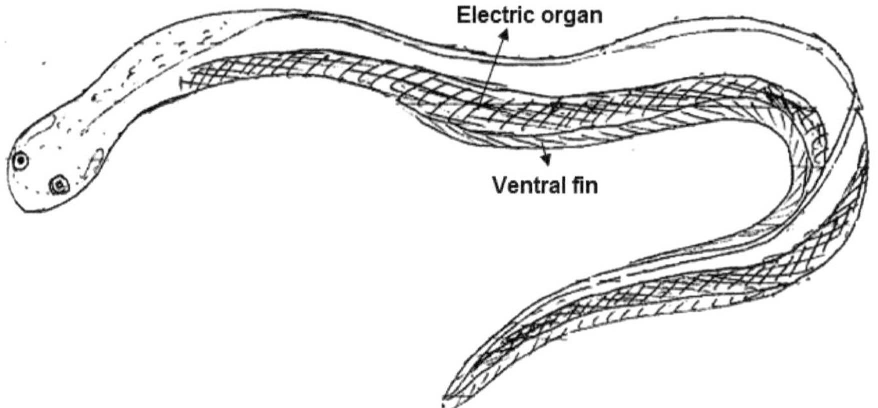

Fig.7.1 : Electric eel.

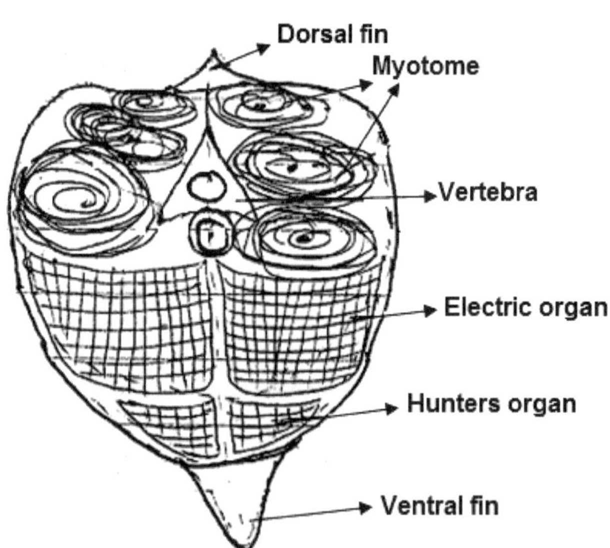

Fig.7.2 : Transverse Section of Electric organ of Electric eel.

The smaller one lying on the ventral side is called the Hunters organ. These organs occupy nearly half the mass of the body and consist of about seventy compartments each containing a series of about 6000 electroplates. The electric organs extend along the whole length of the fish and the nerves supplying them arise from the spinal card. The Electrophorus is capable of discharging a current of about 500 volts on land but in water it is reduced to about 250 volts. The maximum power may exceed 100 watt. It is sufficiently of the organ is form head to tail.

Structure of Electric organs of Electric Catfish

This African catfish (Malapterurus) has peculiar electric organs s it is located in the skin of the fish forming a thick, semi-transparent covering over the body. A number of connective tissue septa divide the electric organ into several compartments fitting into each other. Each compartment contains a large number of electroplates whose nervous ends are towards the tail. The innervations of the electroplates is also very remarkable in this fish as all the electroplates in each half of the organ are connected to the processes of a single nerve cell in the gray matter of the spinal card. The polarity in this fish is from tails to head electric current produced is 350-450volts.

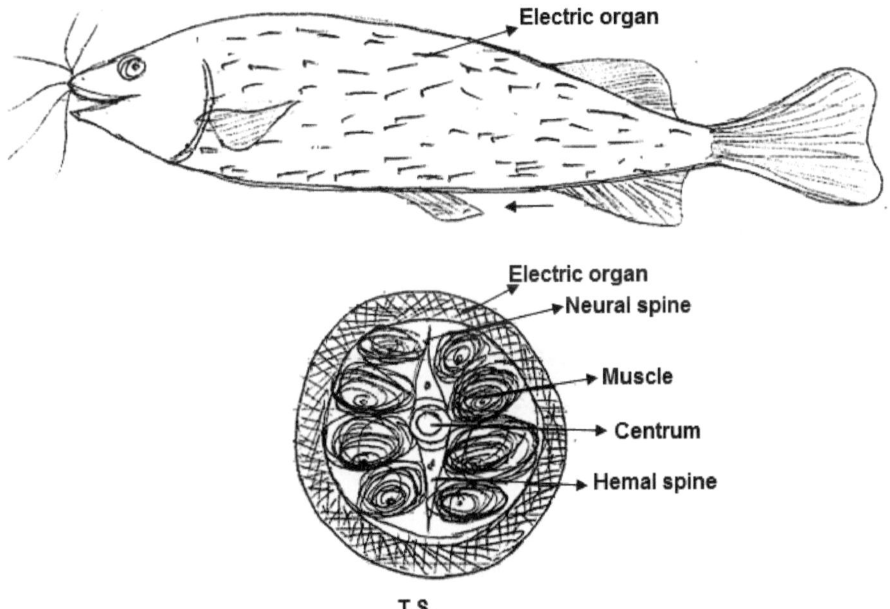

Fig.7.3 : Malapterurus and T.S. of Electric organs of malapterurus

Structure of Electric organs of Electric ray

It is commonly called as Torpedo. The electric ray has two large and two small organs on each side of the head, the smaller one being enclosed within the larger organs. Each organ forms a large flat kidney- shaped mass on either side of the middle line and is supplied by 11 thick nerve originating from a special lobe of the brain. The organ is composed of about 45 vertices hexagons columns each having about 400 electroplated. The innervated side of the electroplate is electrically negative to the other and all work together in parallel. The current in this species passes from the dorsal side (positive) to the ventral side (negative) of the fish. The discharge from torpedo is about 30 volts, sometimes reaching up to 50 volts.

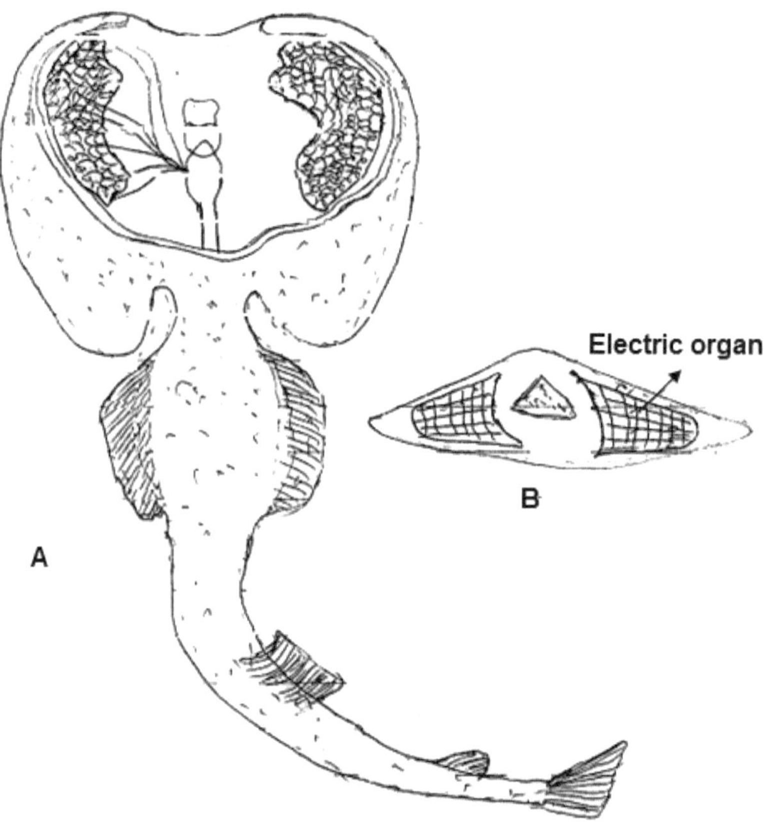

Fig.7.4 : Torpedo and Electric organs of torpedo

Structure of the Electric organs

The shape and position of the electric organs differ greatly in different species of fishes, but all of them have more or less similar microscopic structure.

Each organs is made up of a large number of disc like cells called electro planes or electroplates, that are so arranged as to face the same direction in a given Species. The electroplates are embedded in a jelly like extra cellular material and are bound together by connective tissue into an elongated tube or compartment.

One face of each electroplate is supplied by nerve fibers and the jelly received blood capillaries. Each electroplate is a multinucleate cell with nearly transparent cytoplasm so that the electric organ looks like a clear gelatinous mass, as compared to the muscles.

The electroplate may show characteristic folding and convolutions of one or both of its surface. The havocs end is smooth while the non-heroes face bears large papillae. This is probably an adaptation to improve the efficiency of electric discharge by lowering the effective internal resistance of electroplate due to increate in the total membrane are a differences have been observed in various species in the mode of contact between nerves and electroplate.

Mechanism of electric discharge formation and control of electric organ

The polarity and the strength of the electric current generated by different species varies considerably. The voltage is considerably higher when the fish is in the air than when it is in water. The number of electro plaques in each common and the arrangement of such columns condition the strength of the discharge. The system works on a simple principle of the batteries in series or in parallel. Some of the values of the current produced by fishes are given below :

1.	Skate	4 volts
2.	Ray (Torpedo sp)	37 volts
3.	Ray (Torpedo sp)	40 volts
4.	Stargazer	50 volts
5.	Electric catfish (malaptercs)	350-450 volts
6.	Electric eel (electrophorus sp)	370-550 volts

The study of electric discharge from electro plaque has helped to understand many physiological characteristics of the neuromuscular junction. This discharge from the electric organs in the form of a series of brief pulses and in the species having large and powerful organs, the pulses occur in small group as a reflex response under the influence of the central nervous system. In fishes having smaller organs there is a continuous series of small pulses. It has been shown conclusively that there is a semi permanent polarization of the two ends of the electroplate, the inside being negative. This is due to the difference in the electrolyte concentrations within and outside the cell.

The electric organs in most of the fishes appear to be under the control of the central nervous system. The neurons in torus semi circularizes are reported to integrate lower under phase and probability coded impulses.

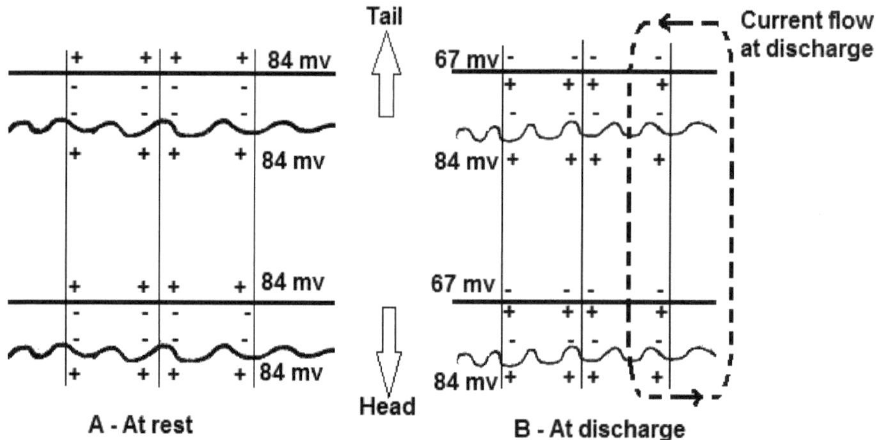

Fig. 7.5 : Mechanism of electric discharge formation

Functions

1. As organs defense and offense: In case of fishes which can generate strong electric discharges often called strongly electric fishes. The electric organ functions as a weapon either defensively against predators a offensively in stunning and capturing food.
2. Electrolocation: The fishes cable of generating low discharges usually described as low voltages of fish weakly electric fish employ the electric organ for electro location, a phenomenon analogous to echolocation in bats.
3. The weak electric current may also help to fish in finding direction in the dark water.
4. Electric discharge may be for maintenance of territoriality by individual fish.
5. Electric organs may also be useful for species even for sex recognition.
6. It is interesting to know that the electric fishes are well protected against their own current and against the discharge of each other.

8

Coloration in Fishes

- Source of colour
- Colour changes in fishes
- Regulation of colour changes
- Significance of colour changes.

The coloration is most well-marked in fishes A large number of fishes are brightly and brilliantly coloured and others are of more uniform and sober shade. On fish body characteristics markings in the form of base strips, spots and blotches etc. coloration is mainly due the skin pigments but the back-ground colour may also be due to the underlying tissues and body fluids. Coloration relates to the life style of the fishes and the patterns so produced have functional significance.

Generally the fishes are darker on their upper surfaces this is because their upper surfaces being exposed to sun light undergo pigmentation , while ventral surface with gradual shading on the sides. Some fish species like the gold fish, carassius exhibit almost uniformly bright colour all over the body. Cave fishes living in total darkness have lost the pigments and grown colourless. Several species of the genera synodontis living in the river of Africa swim in inverted position and so the pigmentation is reversed in them.

The colour pattern in fishes is visible only in the living or freshly killed specimen. After death and in preserved animals the colours quickly fade away.

Bottom dwelling forms are generally marked brilliantly above and pale beneath. A number of freshwater species of fishes are beautifully coloured.

Generally in fishes those side in contact with sunlight that side is show coloured while those side apposite to sunlight is pale coloured.

Source of Colour

Coloration in fishes is due to the presence of two kinds of special cells called as the chromatophores and iridocytes. The chramatophores are branched connective tissue cells present in dermis. Chromatophores contain pigments of various kinds of pigment granules, which are as carotenoids (yellow-red), melanin (black), flavines (yellow), purines (white or silvery), pterins, porphyrins and bile pigments. Depending upon the colour of the pigments the chromatophores are designated as:

Erythrophores	-	Red or orange
Xanthophores	-	Yellow
Melanophores	-	Black
Carotenoides	-	Red, yellow or orange

The chromatophores are of the following types :

1) **Melanophores:** The pigmentary material of chromatophoses is black colored pigment called melanin. Melanin is formed from an amino acid tyrosine which is converted into melanin. A brownish black pigment called eumelanin is also sometimes found within the melanophres.

2) **Iridophores or iridocytes:** The pigmentary material in cells is guanine. Guanine is opaque, whitish or silvery. It is a waste product and is deposited in the form of granules or rounded, polygonal or stellate bodies or in the form of plates. The iridocytes also called as the mirror cells. They give white or silvery appearance to the fish.

3) **Xanthophorcs and erythrophores:** The pigmentary material of these cells include carotenoids and pterins. These white, yellow, orange or red coloured pigments are abundantly found in plants and the fishes being incapable to synthesize them, have to depend upon the plants for a dietary supply of the pigmentary material.

4) **Mixed colouration in fishes:** Colour patterns of a majority of fishes are due to the combined effects of chromatophores containing different kinds of pigmentar granules. Thus the black and yellow chromatophores together develop green colouration. Line wise, yellow and black and orange and blue may combine separately to provide brown appearance.

Colour Changes in Fishes

Many species of fishes are capable of changing their colour or adjusting it to match the surroundings. A rapid adjustment in colour is brought about by rearrangement in pigment granules. The chromatophores undergo concentration or expansion to cause colour change. Concentration means pigments move towards the centre of cell and dispersion means pigments more towards the periphery. Colouration in fishes also varies with age, sex, ill health and emotions.

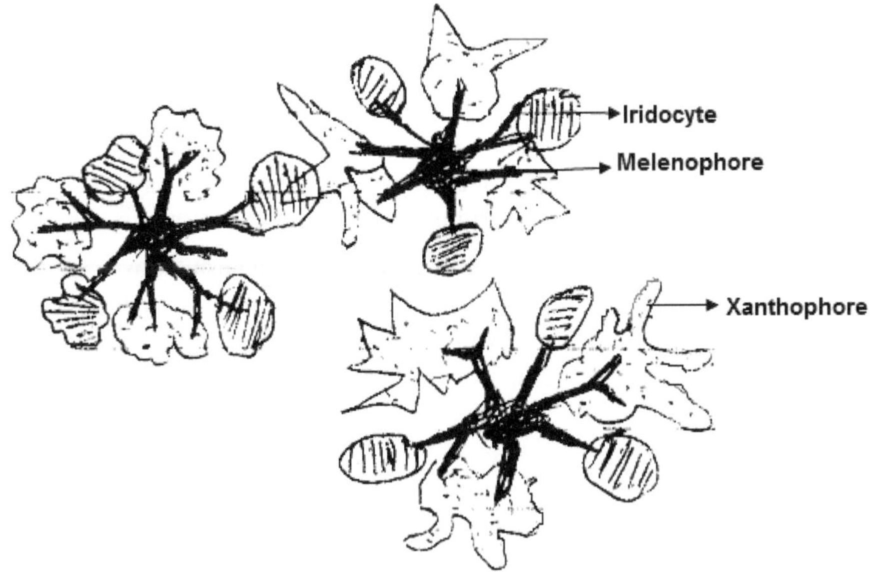

Fig. 8.1 : Chromatophores of the skin of a fish.

In fishes two types of colour changes occurs as below:

Physiological Colour Changes

Fish change the pattern of colouration accordance to their surroundings. The chromatophores may be mono or di, polychromatic and these pigments remain normally concentrated the centre of cells. The pattern of colour changes occurs by the dispersion of such pigments from the perinuclear area to be cellular processes. The dispersion brings about the change of colouration. The rate of physiological colour change is limited. Such colour changes occurs form few seconds as in fundulus several days in Anguilla.

Morphological Colour Change

Morphological colour changes involve changes in the quantity pigments within the fish or its cells. It result in both the crease in the quantity of pigments within each cell and in an increase in the number of functional chromatophores per unit area of the skin, that colour is called as morphological colour changes.

Regulation of Colour Changes in Fishes

Following factors have been associated with the regulation of colour changes in fishes:

i. **Temperature:** The Temperature indirectly regulates the coloration in fishes. The low temperature brings about the darkening effect while it raises result in the concentration of pigments with consequent lighting of the colour. Hence temperature is important in colouration.

ii. **Stimulation:** External stimuli either tactile or psychic influence colouration Tactile to have only little influence on the chromatophores system. Psychic have influence or induces chromatophores system extraordinarily.

iii. **Light:** Light influences the colour change mechanism in two significant ways. The changes brought about by light influencing through other routes instead of eyes such are called as primary responses. Those changes brought about by light through ways of eyes are called secondary responses.

Hormonal Regulation

When in fishes mechanisms to control and co-ordinate colour changes in the chramatophores of body through the hormones are called hormonal regulation. The pituitary gland exercises control over the distribution of pigments in the chromatophores. After hypophysectory the pigments become aggregated and the colour becomes light. If the pituitary extract is injected into a fish temporary dispersion of the pigments is brought about.

It has been suggested that the melanophores of some animals do not respond to the hormones or more than one hormone may be responsible for colour change. The most important pituitary hormone causing colour changes in fishes are Intermedin and MSH (melanophore stimulating hormone). These hormones causes dispersion of the melonophore pigment resulting in the darker colour of the animal, its absence results in the aggregation of the pigment causing their lighter colour.

Another pituitary hormone usually designated as a w-substance is suggested to act an antagonist to the intermedin.

Besides the pituitary hormones, adrenaline also exerts a powerful concentrating effects on the melanophores some fishes show a change in colour during their reproductive period which may be due to gonadal hormones.

Thyroxin produced by the thyroid glands is also reported to bring about colour change in influencing the melanophores.

Neural Regulation

The chromatophores of some fishes are abundantly supplied with nerves. When in fishes mechanism of colour change n the chromatophores through nervous system is called as the neural regulation.

In several fish species chromatophores are greatly affected by cutting the nerves supplying a particular area. It is relived that the neurons supplying the chromatophores produce chemical substances called neurohumors, which activate them. These are two kinds of nerve fibers with opposite effects. One type of fibers produce neurohumers causing pigment dispersion and the other causing aggregation one of these is believed to be acetylcholine and the other is similar to adrenaline.

Significance of Coloration or Colour Changes

The coloration in fishes performs numerous functions and is useful to the fish in a variety of ways the functions are as below :

1] **Camouflage :** Some fishes are permanently coloured to match with the surroundings in which they live. The darker upper and the lighter lower surfaces of a number of fishes are well suited to photogenic conditions of water. Many fishes are able to change their color very fast by rearranging pigments in chromatophores and thus merge with the background and successfully avoid the attention of predators.

2] **Concealment from the enemy for protection:** Various fish species produce patterns to mimic with the surrounding or the fish resembling in color with the surrounding. The fish is also able to change its colour to match the surroundings more perfectly so as to avoid the predators and the prey concealment is achieved by obliterative shading.

3] **Disguise to secure prey:** Disguise is affected by certain habits of the fishes. Disguise means to conceal by changed appearance or to hide with changing colour. Some fishes such as Leather-Jacket fish, Amazon leaf fish these fishes very much resemble a leaf floating in water. Lepisosteus lies motionless near the surface of water imitating a floating twig on the water. Scorpion fish living among the rocks and coral reefs, add weed like tabs to its head to blend with the background for protection and capture of prey.

4] **Advertisement for several purposes:** In some fish species colouration appears to advertise the presence of the fish instead of concealing it. This may be helpful for sexual recognition or it may serve as a warning to the enemy.

5] **Disruptive colouration:** Several species of fishes have developed disruptive colouration as a means of concealment. The colour changes of such kind produce patterns that confuses the form of fish .patches of irregular size, shape and colour draw attention of the observes sight while drawing away their attentions to the forms which bear them.

6] **Aggressive purpose:** The colouration is sometimes for aggressive purpose and not for protection helping the fish to conceal its nearness to the prey. The fish is sluggish in habit and usually hides itself among sea weeds to which it resembles in colour pattern. These may be a device to deceive and attract the prey.

7] **Courting patterns:** Colouration patterns are helpful in several recognition of the male.

8] **Warning patterns:** In some fishes colouration appears to advertise the presence of the fish instead of concealing, it may serve as a warning to the enemy.

9] **Colour patterns in taxonomy:** The colour patterns of the fishes is useful to the fish taxonomist in distinguishing the species and subspecies.

There are variations among the individuals of a species, certain basic colour pattern like, sports, colour bands occur on the body. The location of colour bands or spots or chromatophores is presumably under genetic control and provides a means for identifying the species.

9

Adaptation in Deep Sea Fishes

The oceans covering nearly 71% of earth surface are classified into several zones. There are different zones such as neritic zone, oceanic zone. Neritic zone consists of different zones. Which are present in bank side of ocean. The oceanic zone forms two major zones as pelagic and the benthic zones. In the pelagic zone the organisms float or swim freely on the surface and in the benthic zone they associations are widely different and there is no light, water is called dark zone water or aphotic this constitute the deep sea fishes which found in the bottom of the sea the region between 200-1000 meters is called as the mesopelagic zone. The part of the ocean below 1000 meters is called bathypelagic zone and the deepest part of the deepwater is called benthic zone. Mesopelagic, bathypelagic and benthic zone constitute the deep sea water. A large number of species of fish in habits the deep sea. Fishes living in the deep water of the oceans are exposed to more or less uniform conditions of life and they are adapted differently.

Conditions in the Deep Sea

The organization, habits and physiology of the deep sea fish form is modified to suit the physical, chemical and biological conditions of the deep sea. These are described briefly in the following account.

1. **Physical factors:** The chief physical affecting the fish forma in the deep sea are the temp, light, pressure ocean floor and the currents. The temp of the deep sea is low and is nearly stable between 4 to 8°c. The uniform temp very near the freezing point. The light is completely absent and it is perpetually dark in the deep sea. Absence of light is an extremely important

factor for all life in the sea. Pressure is an important factor influencing life and distribution of fishes. At great depths pressure acts as a limiting factor in the distribution of fishes. The water of the ocean is relatively calm at greater depths and there are no waves or storms except the oceanic currents. The substratum soft and muddy.

2. **Chemical factors:** Salinity and the amount of dissolved oxygen are the main chemical factors that may influence the distribution of fishes. Sea water does not show much variations in salinity and a small differences dose not appears to influence fishes. Dissolved oxygen is very important. Minimum oxygen level exist between 100-1000 m depths and below 1000 m oxygen content is very low which along with the low temp limits the activity of the fishes.

3. **Biological factors:** Predation, competition and scarcity of food constitute the important biological factor. Due to the absence of sunlight plants are entirely absent. Hence food consists of animals and plants falling from above and there is always a shortage of food. Competition for food is therefore an important factor influencing the deep sea community. All deep sea fishes are carnivorel and have developed various structural adaptations. Due to the absence of bacteria and low temp, dead animal do not decompose in the deep sea.

Types of Adaptation

Fishes living in different zones of the deep sea show special adaptations as described below:

1. Mesopelagic fishes

The mesopeligic fishes lead an active life under dim light and undertake extensive vertical migrations, going to the epipelagic zone at night, where they feed on plankton and on each other and return several hundred meters below the surface during the day. These fishes have muscular boding well ossified skeleton, scales, well developed nervous system. Heart and kidneys are large and eyes are large and colour is black. Lantern and stomiatoids are the dominant fishes in this region.

2. Bathypelagic fishes

These fishes show adaptations for a secondary life living in a zone devoid of light excepting the bioluminescence and scarcity of food. They have poorly developed muscles, week skeleton, no scales, poorly developed central nervous system, small gills, small kidneys, eyes are small or non functional, acoustic lateral system and olfactory organs are well developed. The colour of the fish is uniformly black and small photophores are present to attract the prey.

3. Benthic fishes

Fishes inhabiting deep water are similar to mesopelagic ones, with well developed organ system are more diversified. In the bottom of sea or ocean these fishes are found hence these are called as the benthic fishes.

Characteristics of benthic fishes

a) **Abundance:** The number and diversity of deep sea fishes decreases with the increase in depth. This is because the quantity of food gradually diminishes towards the bottom. The part of sea lying under high surface productivity zone contains a large number of large size fishes.

b) **Size and weight:** The size and weight of the fishes also decreases with the increase in depth. This relates to ambient pressure of water column lying above them, availability of food materials and to the problems of sinking. By reducing the size a lesser part of body is exposed to ambient pressure working on it and by reducing their weight the total density of the fish is reduced to counter the gravity. Fishes living in deeper waters have lesser amount of res muscles and so they are less active then those living in depths above them.

c) **Bioluminescence:** The phenomenon of light production is very common among the forms living in bathypelagic or abyssopelagic zones of ocean. Individuals bearing photophores produce light of varied colours which is used for different purposes. It is used differently for the purpose of social signaling, species recognition and in turning the prey. Bioluminescence is also used for indentifying their own species to attract the prey and female for schooling also. This is an adaptation to the absence of light.

d) **Eyes:** Eyes have varied dimensions. The shapes and sizes of the eyes are determined by the depth in which the fishes live eyes are very well developed in some and deep sea fishes. Some fishes have greatly enlarged eyes. The large eyes enable the fish to collect as many as rays of light as possible. Telescopic eyes may be directed forwards as in gigantura. In deep sea fishes the eyes show various adaptations for light perception under conditions of pour illumination. The eyes of mesopelagic fishes are usually larges and adoped to various in dim light. They often have large pupils to absorb a maximum amount of light. The eyes of certain other fishes are more like the concave mireors to maximize the amount of light entering into them.

Bathypelagic fishes living in the complete darkness of the sea, have small eyes but those of benthic fishes vary in size. Many fishes of deep oceans are completely blind. They use a well developed system of lateral line canals; the tectlie organs and other sensory structures to used or compensate the loss of vision.

e) **Food and feeding:** The quantity of food gradually decreases with the increase in depth. Owing to this scarcity of food a tendency towards the strict that fishes are predatory in habit. Many fishes of mesopelagic zone undertake daily vertical migration to surface to feed upon the zooplanktons must of the migratory mesopelagic forms are the predators. They have large jaws, feeding teeth and distensible gut for consuming the prey larges then their own size.

Bathypelagic fishes live in scarcity of food so have a limited budget of energy and reduced organ system. They have well developed jaws and a capacity, to consume large size prey ex. Melanocetus.

i. Bathypelagic fishes depend mostly upon the benthopelagic zooplanktons, many of them feeding upon the decaying ooze have lost the power to masticate the food.

ii. Food material consists of dead animals and plants deep sea fishes consume whatever is available. They prey on one another, and also eat falling excreta of surface animals ex. Chrismous, saccopharynx etc.

f) **Other sense organs:** To compensate for the loss of sight many deep sea fishes have well developed long feelers that keep exploring the surroundings. In some species as Bathypterois has single rays of various fins are longer than the body, acting as feelers Stylophorus paradoxus has a long filament arising from the caudal fin. Such long feelers of deep sea fishes enable them to perceive even the slightest movement of water in their neighborhood, as they are always in danger of being attacked by the enemy. The lateral line system is also very well developed in deep sea fishes and the fish becomes aware of even a slight disturbance in the surrounding water.

g) **Buoyancy regulation :** Fishes of the deep sea are faced with the problems of adjusting to the pressure at different depths, while moving to different areas, sedentary species also have to maintain their position at a particular depth. They requires large amount of energy. Fishes have developed various mechanisms to keep their density as near the sea water as possible such a type of buoyancy called natural buoyancy neither allows a fish to sink or to rise in sea water. It is achieved by one or more different ways.

i. By reducing the size of body, the deep sea fishes reduce the rate of their sinking.

ii. Many fishes by increasing water contents of their body bring their density near to that of supporting water and thus reduce their rate of sinking.

iii. Many fishes attain neutral buoyancy by storing low density lipids. Such lipids have their density lower than the common triglycerides.

iv. By deflating or inflating the swim bladder.

v. Many mesopelagic teleosts use wax esters to attain neutral buoyancy.

vi. Presence of an extra peritoneal air bladder is regarded as an additional mean to achieve neutral buoyancy.

h) Reproductive strategy : Deep sea fishes show remarkable adaptive features for reproduction. These fishes are faced with problems of finding a male due to the absence of light and extremely low density of the species. The mesopleagic and bathypelagic predatory fishes produce epipelagic larvae in large numbers. Ex. Lantern fishes.

The female angler fish has solved the problem of finding a male by releasing a species specific pheromone into the water, which is detected by the male fish. In some fished the male finding the female becomes attached to her means of saws and assumes a parasitic life. In this species go needs of both the sexes do not mature until the female has been parasitized by the male. In different fished the adaptive development to find a male, is to become hermaphrodite, so that any individual can mate with any other individual of the same species.

Some of the deep sea fishes solved the problem of finding a male by producing sound with their swim bladder as in Macrouridae and Ophididae. The sound is believed to attract the male from some distance. Photophores also serve to attract the male some species have developed parental care and few are viviporous also. Thus various strategies have been evolved by the deep sea fishes.

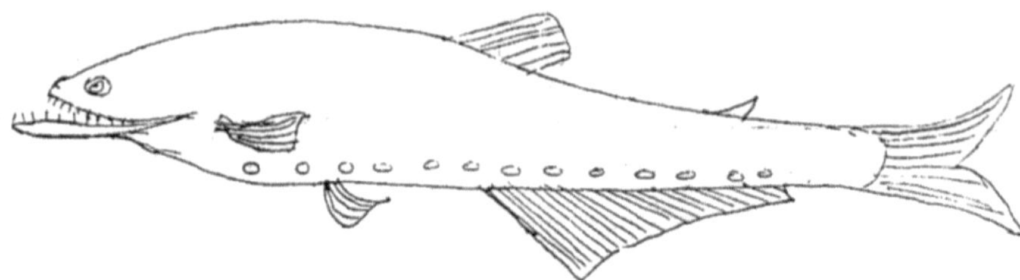

Fig. 9.1 : Bristle math (Cyclothone)

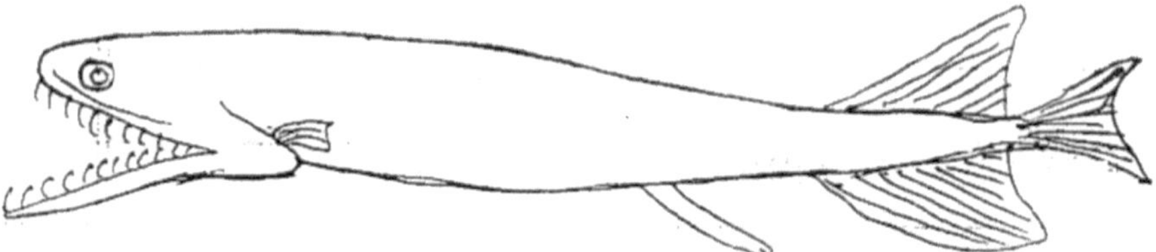

Fig. 9.2 : Loose jaw- malacosteus

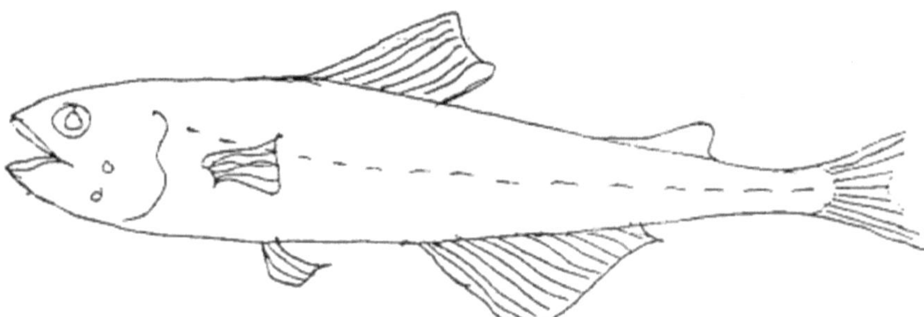

Fig. 9.3 : Lanternfish

Fig. 9.4 : Abyssobrotula

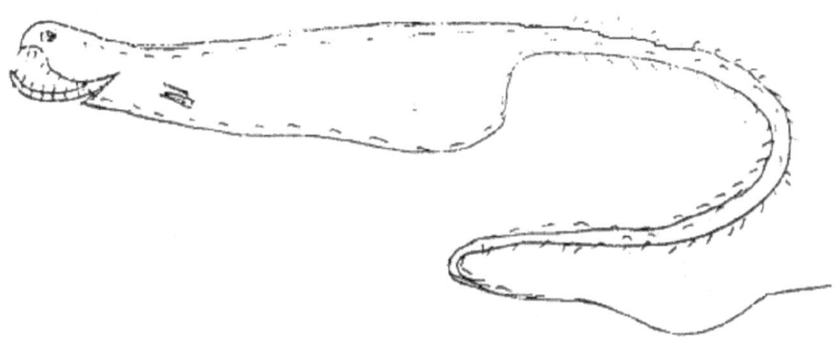

Fig. 9.5 : Gusper eeli

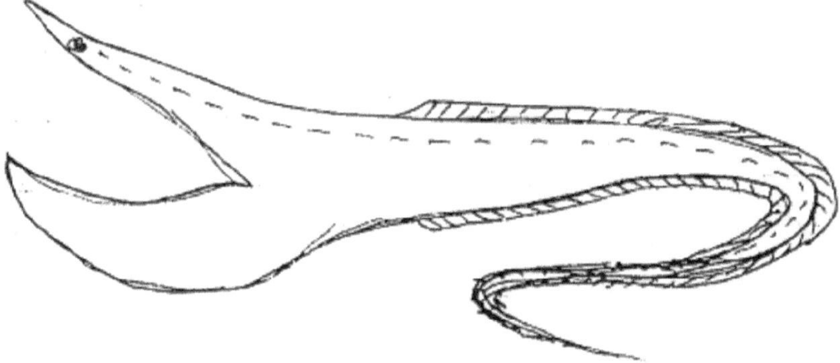

Fig. 9.6 : Gastrostomy

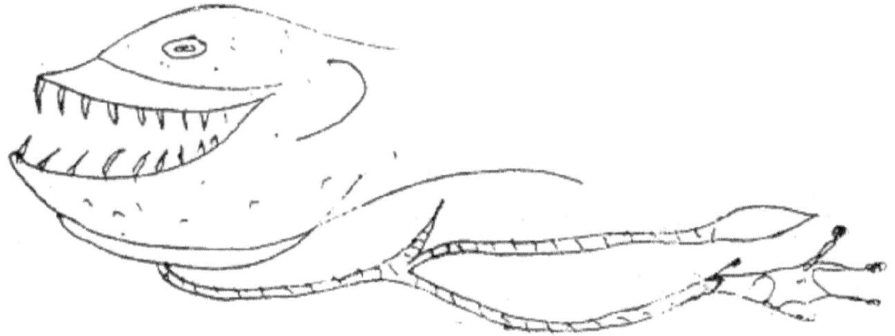

Fig. 9.7 : Deep sea fish with feelers

10

Larvivorous Fishes

(Fish in Relation to Public Health)

Shallow weed infected ponds, swamps, marshes, pools. Bogs infected all inland water bodies. These infected inland water bodies provides shelter food and suitable environmental condition to grow and multiply to different pathogens and microorganism and macro-organism such as mosquitoes, snails other hosts.

Mosquitoes are the vectors of several diseases and carry parasites of malaria, filarial, yellow fever and so on the breed in all sorts of stagnant water and shallow weed infested ponds, swamps, pits gutters and all other kinds of inland water bodies. Different species of mosquitoes prefer different kinds of water for breeding purpose.

A number of fishes have little food value but are of immense utility to public health because of their larvivorous feeding habits. Fish is a natural enemy of mosquito eggs and larvae and its use as means of biological control has been recognized since olden times. Several species of fresh water fishes are known to be larvicidal and of use in the control of mosquito population. Biological control of this type is now regarded as one of the most potent tool in destroying the mosquito population without any causing harmful effect in the food chain and the human system. Man learnt to use them as one of his several measures to combat a number of serious diseases which are spread through the agency of insects. A number of larvicidal fishes directly feed upon mosquito larvae and thus their introduction in such water bodies where mosquitoes breed results in a very effective control on the population the mosquitoes in selection of the larvicidal fishes for such a control of diseases.

A list of various indigenous and exotic larvicidal fishes of India is important. These fishes possess the different specialized characters. Due to these characters these fishes are notable and famous and called as larvivorous fishes. These characters are as below :

Characters / Features of Larvivorous

1. It should be of small size. That enables the fish to move freely among the shallow at water weeds in hunting for larvae.
2. It should be a hardly fish with respect to the ecological conditions as surviving in deep as well as shallow water and to withstand transport to long distance.
3. It should breed freely in confined water i.e. naturally spawning under ordinary environmental condition.
4. It should active and fast runner and swift enough to escape from the enemies.
5. It should be a surface feeder and carnivorous in habit.
6. Adaptations in jaw mechanism to capture and feed upon the larvae.
7. It should be have no food value or little food value, though of immense value as larvivorous fishes.
8. It should be difficult to catch.

With all such characteristics the fishes mat be used as substitute of insecticides in carring out the biological control of vector insects.

The eggs, larvae and adult mosquitoes form an excellent food for several species of fish, but those of small size prefer mosquito larvae and feed on them throughout their life. These species are more useful as larvicidal fishes.

There are several exotic and indigenous species which at one stage or the other feed on larvae and are considered as larvicidal fishes.

Exotic species –	Gambusia affinis – Guppy
	Lebistes reticulatus
	Catassius auratus – Goldfish
Indigenous fish -	Notopterus
	Oxygaster (Chela)
	Danio
	Rasbora daniconius

Amblypharyngadon mola

Anabus

Mugil

Puntius

Wallago attu

Channa sp.

Barilius

Ambassis

Etroplus

Aplocheilus etc.

From the point of view of mosquitocidid activity the above stated and other probable larvicidal fishes have been classified in to the following groups, according to their efficiency in mosquito control which are as follows :

1. **Typical surface feeder:** These fishes satisfy all the required conditions of mosquitocidal fishes, hence considered excellent for mosquito control. These species are generally surface feeder. These mosquito larvae and nymph are present on surface of water, which can be easily consumed. Example – Gambusia, Aplocheilus, Lebistes, Aphanius etc. Species like Oryzios, Lebistes, Aphanius are also surface feeder but are less efficient due to their mode of life.

2. **Subsurface feeder:** Subsurface feeder like Amblypharyngodan, Danio, Rashora, Esomus and carassius, which are larvicidal to a considerable extent. They are easy to raise but considered less efficient larvicidal fishes.

3. **Column feeding:** These fishes which swim in column water area these are called column feeders. These fishes include the puntius Ambasis, Anabus, colisa, Therapon etc. are effective larvicidal forms but they feed on larvae only by chance.

4. **Large sized food fishes:** Those fished which are attaining a large size and used as foods for human. Hence it is called large sized food fishes such as catla, Labeo and mugli are also helpful in reducing the larvae of mosquitoes because their fry feed chiefly upon the larvae and planktons.

5. **Predatory fishcs:** The fry of many predatory fishes i.e. wallago, Channa and Notopterus consume mosquito larvae as these chief diet and help in

controlling the population of mosquito but the young and adult fish are destructive for other fishes.

The larvicidal fishes are introduced into the water which are the chief mosquito breeding sites. The water is cleared of predatory fishes and dense vegetation.

The utility of controlling mosquito through fish is limited to permanent and semi-permanent pieces of water. Areas which are inaccessible to fish or temporary collections of water from suitable and safe place for breeding of mosquitoes. Hence the use of larvivorus fishes alone cannot solve the problem and has to be supplemented by other methods.

As in the mosquito control , one of the effective method to control guinea worm is to introduce cyclopsivorous fishes such as Ambassis in order to cut down the population of cyslops. Many species of fish feed on offal and filth and help to reduce organic refuse which would otherwise accumulate putrefy and pollute the water.

The control of emergent and submerged aquatic vegetation helps to check mosquito population.

India conducted National Malarial Education Programme (NMEP) in 1958 designed to interrupt the life cycle of the malarial parasite with insecticide spraying chemotherapy in affected human and mosquito larvae control methods in rural areas. The incidence of malaria dropped from 75 million cases at 100000 per year.

In 1976, 7.4 million parasite positive cases are reported by resources as:

1. DDT shortage throughout the period of the project operation extent of 13-34%.
2. Persistent malaria foci.
3. Continuation of urban transmission.
4. Poor surveillance.
5. Physiological resistance of the vectors to insecticides used at the time.
6. Drug resistance by the parasite.
7. Resistance to spraying by man.

While the NMEP was a vertical centrally administered approach offer control failures in the 1970 a revised strategy called "Modified Operation of Plan" (MOP) was adopted the objectives of which were to control morbidity and mortality rather than aim at eradication of malaria.

By 1978 however following India's commitment at the Alma Ata conference to achieve "Health for All by the year 2000". The programme adopted a national directive to integrate anti-malaria operation with Primary Health care services is PHC develops. All these aim is to eradicate the larvae and mosquito.

Following is the brief account of same larvivorous fishes :

1. **Exotic fishes:** There are three exotic species which are larvivorous such as carassius aurotur or gold fish Lebistes and Gambusia. The Gambusia or top minnow or Guppy has been extensively used for mosquito control. Gambusia is very voracious and destructive fish for mosquito larvae and mosquito also.

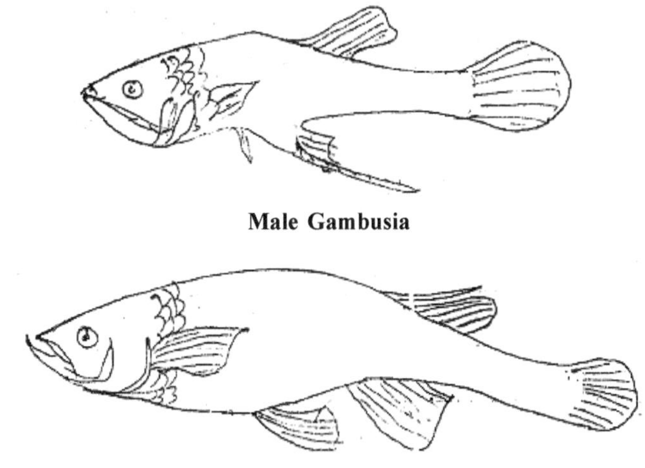

Male Gambusia

Female Gambusia

Fig. 10.1 : Male and female types of Gambusia fishes

2. **Notopterus notopterus:** This fish is found in fresh and brackish water of India. The young stages of this fish feed on mosquito larvae but the adult is feed on small fish and fingerlings so it is harmful it cannot be recommended for anti malaria use.

3. **Rasbora daniconius:** This small sized fish is found through out in fresh water of India. It is hardly and efficient larvicidal fish, through it is also consumed as food by poor sections of the people. It can be used in regions where it occurs abundantly.

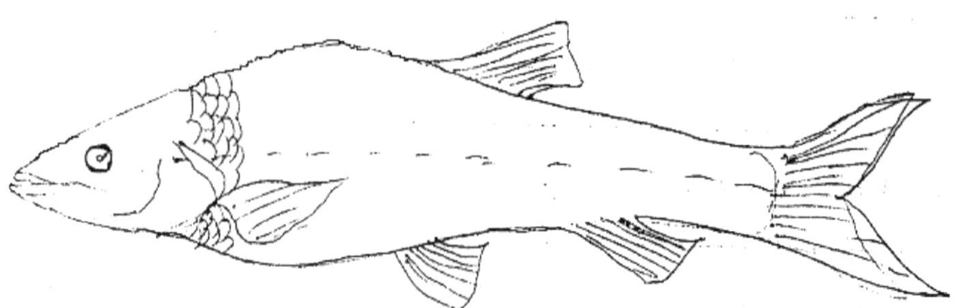

Fig. 10.2 : Rasbora Fish

4. **Puntius :** This fish is represented by several species found through out in the fresh waters of India. They are popularly called carp minnows they are small, hardly species and are consoled locally by poor people. They can withstand transport and breed in confined waters. They are useful as larvicidal forms. Two black spots are present on the body one on behind opercula and other on base of cerebral peduncles.

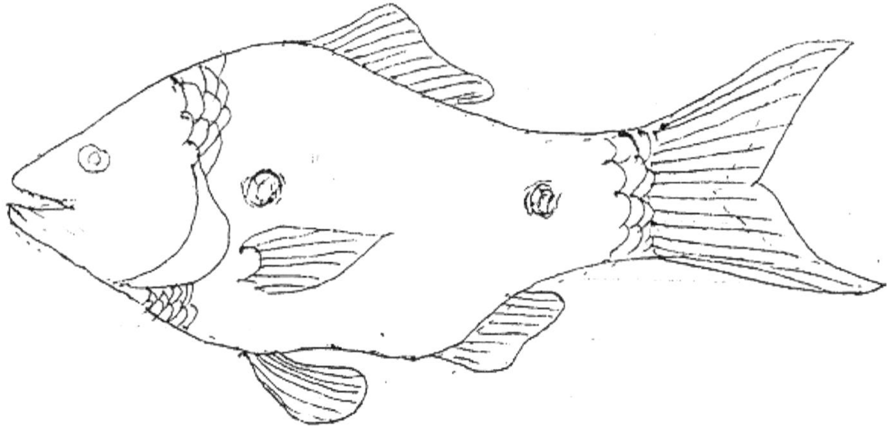

Fig. 10.3 : Puntius fish

All major carps as catta, Labeo, cirrhina feed on insects and crustaceans in their young stages. Hence crap cultural is useful for mosquito control as well as for providing food fishes but these species cannot be called as larvicidal forms.

11

Aquatic Weeds and Their Control

- Type of Aquatic weeds
- Different control method
- Advantage and disadvantage of Aquatic weeds.

Aquatic weeds are those unwanted and undesirable plants, which reproduce and grow in fast in water bodies. They are more harmful them beneficial for fish culture. Excessive growth of aquatic vegetation prevents effective utilisation of water and reduces productivity because they utilize the nutrients and also prevent the sunlight to read the deep part of the water body. This excessive growth restrict the living space of fishes, resulting into less production. The decay mass of these pollute weeds prevent the growth of more useful microorganisms. And disturb the food web and food chain and equilibrium of physico-chemical qualities of water. Liberation of toxic substance and waste decomposition of water body is most effective so it is necessary to identify and control them. It has the primary problem of management of Inland fishery cultures.

Aquatic plants grow on extremely wet soil and where water is available to plants in abundance such as ponds lanes river banks etc. or those plants which grows and found in the nearby water body or in water body. These plants are called as aquatic weeds because these are not attend the larger size and act as the weeds which is found in water so these are called as aquatic weeds.

According to their distribution in a fish water body following different types of aquatic weeds have been recognized.

Types of Aquatic Weeds

Aquatic weeds can be classified into following groups:

1) Free floating or surface floating weeds

They are not attached to bottom sediment. They drift in the water along with the water current. They float freely on the water surface. They prefer quiet water for their optimum development. They remain in contact with water and air but not soil.

They float on the water surface with most of their body above the water surface of the roots are present they hang in the water and are not anchored to the bottom. These are designated by the symbol "T"

Ex. Lemna, pistia, woffia, spirodella eichornia etc.

Fig. 11.1 : Pistia weed

2) Submerged floating weeds/free floating submerged weeds

Some of the prominent submerged floating plants they spend their entire life cycle submerged in the water. They have no toots. Their flowers are elevated above the water for insect or wind pollination except in ceratophyllum where they are submerged and have hygrophilous pollination. The vegetative reproduction is more common in these plants. They perish when exposed to strong illumination the transpiration stream has no existence in submerged weeds or plants.

Thin branched stems totally devoid of roots and covered with delicate hair line simple or compound leaves divided into hair like segments, float just beneath the water surface. Leaves bear ting bladders which act as traps for small aquatic invertebrates which serve as food for the plant. Bladders one interpreted as a modification of part of the leaf, flowers are yellow purple with one or several born on a thin wiry scope that protrudes in the air and entomophilies pollinations.

Ex- Utricularia, ceratophyllum, aidrovandies etc.

Fig. 11.2 : Ceratophyllum and Utricularia

3) Attached plants/ weeds

They grow in shallow water up to depth of effective light penetration

Along the water margin most of them favour and or mud-sand substratum for their grow and attachment and are not collected from rocky substratum. Their roots functions more as an organ of anchorage to mud and erectness of flowering stem than for the absorption of water and nutrients these are as follows :

a) **Rooted weeds with floating lives:** In this group of aquatic weeds the roots are fixed in mud but leaves have long petioles which keep them bloating on the water surface. Their horizontal photosynthetic leaf lamina floats on the water surface and they more effectively reduce the light penetration in the water than other surface floating panty and algae. Their leaf bides are usually circular and have a tough leathery tax use. That offered protection against tearing and perforation. A leaf with an entire out line is less easily watered and submerged than the one which is subdivided is obvious that the centre of gravity of a floating leaf lies at it is central point and that this is therefore the most mechanically economical position for petioles support. The upper surface of the leaf is usually covered with a waxy cuticle but the lower surface is devoid of it.

b) **Root of the plant body is submerged in water:** Ex. Nymphea, Nelumbo, brasenia, trata, marsilea and aponogeton etc.

Fig.11.3. Trapa and Nymphea weeds.

4) Rooted plants with submerged leaves

Some aquatic weeds remain completely submerged in water but rooted in the soil. The majority of them are characterised by thin elongated, branched stem clothed with leaves and often capable of rooting at the nodes. The submerged stem or eaves. Have little or no cuticle on the surface and facilitate the dissolved of dissolved gases and nutrients in to and out of the tissue.

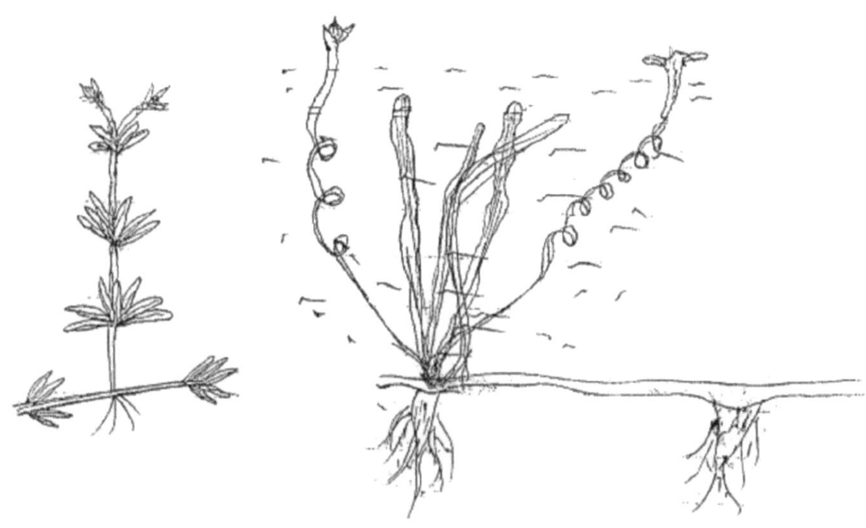

Fig. 11.4 : Hydrilla and Vallisneria weeds.

The submerged leaves in general are thin usually long either entire or deeply dissected and efficient absorption of dissolved gases and nutrients from water. The submerged leaves are poor in stomata and hair.

Ex. Hydrilla, potamogeton vallisneria, chara, najus isotes etc.

5) Rooted plants with emergent leaves /Rooted emergent

These plants grow in shallow waters and require excess of water but this shoots are partly or completely exposed to air. The roots system is completely under water, fixed in soil, these are called as emergent weeds.

They are rooted plants with principal photosynthetic surfaces projecting in the air above the water. They obtain carbon dioxide from the air but nutrients from beneath the water surface. An anchorage of plants to bottom midair sand prevent from being swept away by water current.

In some emergent weeds shoots are partly in water and partly exposed all of water in air example- Ranunculus, sagitari limnophilia and monochoria. In some emergent weeds shoots are completely exposed to air like the land plants called as marsh plants ex. Typha cyperus, Rumex etc.

Fig. 11.5 : Sagitaria and Typha weeds.

Different Methods of Weed Controls

There are three different types of aquatic weed controls which are as follows:

A. Mechanical control.

B. Chemical control.

C. Biological control.

Lawrence (1949) checked the aquatic weeds and establishing them as given below:

1. Keep minimum depth of water at not less than 0.5 m.
2. The pond edge to reduce hard than shallow water weeds growth.
3. The shape and shading of the pond edge above water level to reduce the area, the marginal weeds appear.
4. Muddy area is required around the pond.

A) Mechanical control

Mechanical removal of weeds by employing human labour or by machines is the most satisfactory method. Where the labour is cheap, manual labour is often employed periodical removal of aquatic plants by hand picking , uprooting or cutting, keeps them in check. The rooted submerged woods are removed by dragging log welders fitted with spikes and barbed wires or by cutting them with long handled forks and stickle.

During the recent year power lifts or diesel operated machines have been used for eradicating dense rooted vegetation several most common floating weeds are found in lakes and pond. Complete removal of these weeds by manual labour, drying and burning it up is considered the most effective method of eradicating it. The mechanical method of weeds in India is of limited scale. The method is slow, rather expensive and generally pisciculturist adopt it. In Bengal new method applied for cutting the weeds long "v" shaped sickles are used for cutting weeds.

There are also other useful method which may be chracterised managed such as control of marginal weeds and grasses, control the algal blooms and nursery pond by shading.

B) Chemical control

Several chemical weedicides are now available for the control of aquatic weeds, but they have to be used carefully to prevent adverse effect on the fishes in water. The chemical is to be selected and used in such a way that.

i) It should be cheap and easily available.

ii) Non toxic to fish and man.

iii) Should not pollute the water and

iv) Should not involve the use of special and costly equipment.

a) Control of floating weeds:

It can be successfully controlled by the chemical 2,4 dichlorophenoxyl acetic acid (2,4,D). It applied at the rate of 4.5-6.5 kg/ha and has little harmful effects on fish, it may absorb root and dead it. Another acid weedicide is Taficide -80 (2,4 D sodium salt 80%) used at a rate of 4-6 kg/ha in an aqueous solution of 1-1.5 concentration in combination with 0.25% solution of surf is important. Simazine is also an effective weedicide.

3.5 kg chloroxone in 500-1000 liter solution sprayed on weeds it will kill within 2-3 weeks. 40 ppm of ammonia is also effective.

b) Control of marginal weeds:

If we use the 4D-sodium and 2-2 dichloroproponic sodium aqueous solution at the rate of 10 t0 12 kg/ha.

1-1.5% aqueous solution of Taficide-80 along with detergent surf is best effective.

c) Control of emergent weeds:

Use of Taficide-80 is reported to be effective in controlling weeds. At the CIFRI, marginal weeds like Nymphoides and Nymphea have been eradicated to some extent by spreading copper sulphate with mud on the bottom soil by 4-5 intermittent doses.

d) Control of submerged weeds:

To control submerged weeds copper sulphate in combination with ammonium salt is also found to effective. Sodium arsenate at 5-6 ppm is also found to effective Anhydrous ammonia is also effective control weeds.

C) Biological control

Certain type of weeds can be controlled by means of selected varities of herbivorous fishes such as grass carp, Cyprians carpio or common carp, Tilapia mossambiea , gourami etc. The important herbivorous species and effectively keeps certain aquatic weeds. It is easy to handle, does not biologically interfere with other fishes. The grass carp is an important herbivorous species and effectively keeps certain weeds in check. Advanced pry and fingerlings consume advanced fingerlings, juveniles and adults feed an Lemna Azolla and Salvinia.

Submerged weeds like Hydrilla, Njas ceratophyllum is consumed by the grass carp . In china the grass carp is reported to consume 40-70% of its own weight of grass/day and 100 fish ha can totally eradicate weeds.

Tilapia and Gourami are also found to be great help in keeping the submerged weeds under control – Beside fishes other animals like ducks and geese can also be employed for weed control.

According to the observations of Alikunhi and Sukumaran (1964) the grass carp stocked at 300-375 by numbers weighting 78.8 to 173 kg/ha cleared a pond checked with hydrilla within a month. The fish consumed weeds at the rate of 1.8 kg/day grass.

Interesting observation find out the utilization of aquatic weeds can be controlled by grass carp. It is most preferred food of fish.

In Indian ponds heavily stocked with Tilapla masambica have been found generally free from soft submerged vegetation and filamentous algae.

Submerged aquatic weeds can be destroyed to a limited extent by shading the pond using surface cover of plants.

Inorganic fertilizers mixed with 10% sodium nitrate are used for controlling submerged weeds. Biological control by using herbivorous fish is probably the best method for controlling weeds. Once the weeds eradicated, proper maintenance of the pond to prevent regeneration and growth is necessary, so that the pond is best utilised for fish culture.

Advantages and Disadvantages of Aquatic Weeds

a) Advantages

Aquatic plants or weeds, when in limited quantity are useful and necessary for the ecology of the pond. They form natural food of many species of fish and fertilize the pond when decayed. They provide shade and shelter to many fish and oxygenate the water. They reduce turbidity and provide spawning beds for fishes.

b) Disadvantages

Aquatic weeds are harmful as they consume nutrients of the pond and obstruct netting operations. They check free movement of the fish and cause oxygen depletion and accumulation of carbon dioxide. Gases like hydrogen sulphide and methane are formed which are harmful to the fish. Algal blooms choke the gills and spoil the water on rotting.

12

Composite Fish Farming in India

In order to obtain maximum yield of fish from a lake or reservoir, it is essential to culture fast growing compatible species of different feeding habits is called as composite fish culture.

Any one kind of species cannot utilize the variety of food available in the pond. For a complete utilization of cheap natural food, pond fishes of different age groups or of different species may be selected for stocking. A combination of several compatible species with complementary food habits would make better use of the food resources of the pond as well as ecological potentials of the pond.

When fishes of different age groups and different species are cultured in same pond called as composite fish farming or mixed farming or polyculture. This is based on the principle that compatible species do not harm each other instead they utilize in the most efficient manner ,all available food supply of the pond for maximum production of fish yield. There is no competition between different species on the other land they may be beneficial effect on the growth of others.

The principle requirements of the different species in combination are the following :

1) They have complementary feeding habits.
2) They occupy different ecological niches.
3) They attain adult marketable size at the same time.
4) They should tolerate each other.
5) They should all be non predatory.

Composite carp culture in India is an age old practice. The fishes like catla

catla as a surface feeder. Labeo rohita as a column feeder and Cirrhinus marigla bottom feeder when stocked in a ratio of 3:3:4 utilize food from all the levels of the ponds.

Mixed farming involves rational manuaring and fertilization of pond as well as feeding the fish supplementary food consisting of oil cakes and rice or wheat brine. According to Alkunhi 1957, catal, rohu and mrigal are stocked in the ratio of 3:3:4 to give good yield from the ponds.

Culture practices are not restricted to only the ancestral combination but many more combinations were formed and proved a success. Hora and Pillay (1962) proposed culture of catla, rohu, mrigal and calbasu in proportion of 39:50:10: 10.

Composite fish culture at cutfack invoived a maximum of 7 species i.e. grass carp, silver carp catla, rohu, scale carp, mrigal carp and tilapia in the proportion of 25:15:15:30:40:10:15 respectively. Many more experiments conducted at Central Inland Fisheries Research Institute (CIFRI) and by Indian Council of Agricultural Research (ICAR) have indicated possibilities of high yield and income .

Culturing of a single species is not profitable one does not get the most from that particular water body. Therefore a group of fishes are selected, each having different feeding habits from others so that all the food available in the different zones of the pond is used profitably .

Stocking of the three Indian major carps is catla, rohu and mrigal, in the same pond is an excellent example of the correct selection of species for the maximum utilization of the food from different zones.

Catla is the surface feeder feeds mainly on the zooplanktons and detritus. Mrigal and labeo calbusu are bottom feeders feeds mainly on decaying vegetation, benthic animals plants and epiphytic planktons and rohu is a column feeder and feeds on column food materials. All these give a best yield in ponds.

Exotic carps such as silver carp, grass carp, and cyprinus carp have recently been selected for culturing along with the three Indian major carps with remarkable success. Silvercarp is surface feeder feeds only on phytoplankton unlike catla which feeds mainly on zooplankton. The common carp is an omnivorous bottom feeders utilizing mainly the food which mrigala is not able to consume. The grass carp is known for its efficient utilization of all noxious and excessive macro-vegetation and occupies an ecological niche not utilized by any other fish .

Various combinations selected for experimentation chiefly include:

1) Grass carp silver carp, scale carp and minor carp in ratio of 5:3 :8 :2.
2) Grass carp, silver carp, rohu, scale carp and minor carp in ratio of 5:3:6:8:2
3) Grass crop, Silver carp, catla, rohu, scale carp, minor carp in ratio of 5:3:3:6:8:2
4) Grass carp silver crap catla rohu Scale carp miner crap x Tila 5:3:3:6:8:2:3

Following some factors are considered to useful for full utilization of natural of supplementary food and higher productivity is formed.

1) Selection of suitable combinations with respect to food and feeding habits and their tolerance limit to various abiotic and biotic factors .
2) Stocking densities of fingerlings stocked area of the pond.
3) Initial total weight of fingering being released in the pond .
4) Rate, kind and frequency of fertilization of pond .
5) Rate, kind and frequency of artificial food being supplied.
6) The physic-chemical and biological characteristics of the pond.
7) Fish production includes the grass weight of the fish harvested/ha area of the pond.
8) The cost of production, based on expenditures incurred and the net profit gained.

Composite fish culture also have a great scope in upland water masses of India. Several experiment conducted during the recent years have shown that the production of fish can be increased by 8 to 9 times it the following indigenous and exotic are cultured together as catla, rohu mrigal silver carp, grass carp and cyprinus carps.

During the recent years ,composite fish culture has been taken up in large sized ponds as at the Anjana fish farm in west Bengal farmer average production of 4000 kg/ha/yr. In experiments conducted at the pond culture of the CIFRI at Cuttack and record the yield of more than 9000 kg/ha/year.

Experiments have also been conducted on integration of aquaculture with the livestock rearing to reduce the cost of fish production. In a fish cum-duck rearing center an average yield of 4323 kg/ha/year was obtained without adding any fertilizer the pond and giving any supplementary feed to fish.

In fish-cum-pig culture yield was 7300 kg/ha/year when pig dung 2000 kg/ha was recycled through the pond. Prawn culture in combination with major carps has also been successful.

Among the other fish species which can be reared with the carps preferred ones are the predatory cat fishes and murrells. By applying modern methods of fish farming and using suitable combinations of fish species for composite culture has an aquaplosion or required abundant fish production can be Achieved in our country.

In all the instances of composite fish culture and particularly in the additional fish polyculture or mixed culture careful observation is needed throughout the period of their vegetative growth.

A wider introduction of mixed culture composite culture practice of pond fisheries at fish farm would make it possible not only to increase the quantity of pond fishes and productivity but would also ensure higher income with lower investments.

Most of the investigators have used hormones to include egg and sperm production in fish but the precocious sexual maturation and the inhibition of gonadal development is prevented, all food will be converted into fish and there will be no wastage of food by its converted into gonadal material. Hence production of asexual or sterile fish would be of advantage in commercial fisheries.

When all above different fish species were cultured carefully and forms a best result. For good and best result the following observations were made on composite farming:

a) When the Indian major carps alone were cultured the average yield was 2088 kg/ha year.

b) When exotic carps alone were cultured, the average yield was 2900 kg/ha year.

c) A combination of Indian and exotic carps gave an average yield of 3085 kg/ha year.

Hence above three points, the thin plant is important when Indian major carp, exotic major carp and other species of fishes cultured in combined in a pond then it will gave a best result with the help of composite fish farmings.

13

Economic Importance of Fishes

Fishes are used by human beings in different forms from time immemorial millions of human beings suffer due to hunger and malnutrition and fishes from a rich source of food, which has following importance:

1) Fish is an excellent food for mankind is more easily perishable than cattle, sheep and chicken.
2) Fish flesh is easily digestible by child and man.
3) Fish flesh has very great nutritive and medicinal values.
4) Fish flesh contains proteins, bass, vitamins, waters and minerals.
5) All the essential required amino acids are present in the fish flesh.
6) Fish flesh contains proteins about 13-20 % and it is easily digestible and nutritional.
7) Fish flesh is highly perishable commodity constituted as 60-80 % water.
8) Fish flesh contains about 1-8 % fats or oils and it is more important value of fishes.
9) Fish flesh contains important minerals such as Calcium, Potassium, Phosphates, Iron, Copper, Zink, Magnesium, Iodine, etc.
10) Fish flesh contains a very high percentage of vitamins such as A, D, E and K.
11) A special feature of fish flesh food is contents of vit_{12} which is almost absent in plant food.
12) Certain by product of fish industry are used in various ways.
13) Certain by product of fish industry are used in Ayurvedic and Unani system of medicine.

14) Fish by-products such as fish liver oil and body oil is used in various industries.

15) Certain by – products are used in night blindness, skin diseases, cold, cough, bronchitis, asthmas and tuberculosis

16) Surplus quantity of fish is used as the food of cattle and poultry, the food of Cattle and poultry is called as fish meal which is very high nutritive values.

17) Fish that are unfit for human consumption are used to prepare fish manure Plantations such as sugarcane, tobacco, coffee and coconut etc.

18) From waste part of fish such as gill, bladder is used to prepare fish glue.

19) The skin of several fishes like the sharks and rays are used for preparing ladies shoes, money bags, suitcase, and belts.

20) Fish scales are useful for manufacturing the artificial pearls,

21) Fish flesh is used as nutritive products as fish wafers, fish biscuits are.

22) Some fishes are used for biological control such as malaria.

Medicinal Values of Fishes

Fishes have great medicinal values which are as :

1) The sharks have been found valuable for extracting the insulin form their pancreas.

2) It has been recorded through ethnozoological observations stories or otoliths are taken out and after rubbing and mixing them with water they are given to convalescent children suffering from rickets.

3) Otoliths are fine concretions of calcium carbonate useful for proper growth of bones.

4) It is again from the ethnozoological records that amphibious cuchia has been highly medicinal fish.

5) As per the statements by the fishes folks as soon as the fish is caught, gram flour or wheat flour is rubbed on its body so as to wipe out all its mucous along with flour. Now the flour moistened with mucous is made into small balls and then dried up. Such balls or tablets are prescribed to the persons suffering from impotency and it is claimed after a limited duration course of treatment people regain then loss vigor and strength.

6) Its flesh on the other hand is also considered of high nutritive and medicinal value to prescribe to convalescent people.

7) Among live fishes such as *Clarius batrachus H. fossilis* and *channa* sp. are well known for their highly nutritive and medicinal properties and generally fetch relatively 'high price'.

8) Though Notopterus is relished both in fresh and dried state the soup made from it is reported to be given in measles.

9) Sillago sihama is an excellent food and considered well nourishing for nursing Mothers.

14

Fish By-products

Fish are of great economic value for man. In addition to it is an important item of food and provides several important by – products also which are as :

1) Fish oil

The most important fish by product industry is fish oil which acts as a vehicle for fat soluble vitamins i.e. A, D, E and K, and source of essential faty acids.

The oil of fish is categorized into two main types as the fish liver oil and fish body oil.

The oil abstained from the whole body of fish is called as body oil and is poor in vitamin contents.

The oil abstained from the only liver of the fish is called as the liver oil, and is good quantity of vitamins.

The body oil is less valuable as compared to the liver of the fish.

2) Fish manure

Surplus fishes or that unfit for human consumption are converted into fish manures. Fish manures are used as fertilizers for money plantations as coffee, tea, tobacco, rubber, sugarcane etc. Fish manure prepared from the dried and putrid fishes. Three kinds of fish manures as fish manure, prawn manure and fish guano.

Fish guano is prepared from the fish materials left after the extraction of oil. It contains 7-10% nitrogen and phosphates and considered a rich nutrient for the plants.

3) Fish meal or Fish as food for cattle

Fish meal is defined as a dried fish that has been ground and used as an ingredient of food or as a fertilizers is called fish meal.

Fish meal is prepared from waste fisher left over after extracting oil from the fish. It is also prepared from non edible fishes of both the small and large sizes. Fish meal is also rich in fat soluble vitamins like A, D and K along with water soluble vitamin - B and B12

Fish meal constitutes a valuable source of food for pig, poultry and cattle.

4) Fish flour

It is a high quality fish meal but farms an ideal protein supplement to human diets. It can be easily blended with wheat or maize flour and issued as an enriching component in bread, biscuits, cakes, soups and gruel. It is a valuable food for convalescents and for those suffering from the nutritional deficiencies.

5) Isinglass

Isinglass is a gelatin like material obtained from inner lining of swim bladder which when put in water swells up but does not dissolve in it. It is a high grade collagen. It is produced from air bladder or swim bladder of catfishes, carps, cods etc. Isinglass is used principally for the clarification of wines, beer and vinegar. To a limited extent it is used for preparing jellies and special cement. It was used as a substitute of gelatin in confectionery.

6) Fish glue

Fish glue is a liquid and is used as an adhesive for stamps labels etc. It is also used for making paper beers, shoes, and furniture. It is prepared from bones, skin and connective tissue of fish.

7) Fish leather

The skin of several fish species such as the cod, sharks and rays is used for making leather. Fish leather used for making shoes, bags, suitcase and other ornamental articles.

8) Fish silage

This is a good animal food and is either liquid or semi solid product of a high nutritive value. These fishes are not consumed by man these fishes are minced and mixed with dilute sculpture acid and formic acid and product is used as animal feed. Silage means fodder preserved in a silo.

9) Fish pearls

The silvery scales of cyprinoid fishes have been extensively used in manufacturing of artificial pearls

Artificial pearls are prepared from hollow glass brads which are polished with scales. The hollow brads are then filled with wax and are used in jewelry.

10) Fish soup

The fins of sharks are dried and exported to other country like china where they are used for the preparation of soups. Shark fins have been a fancy for preparing fin soup.

11) Fish insulin

The large sized fishes are dissected so as to remove the pancreas for obtaining insulin. Pancreas of the sharks is rich in insulin.

12) Fish skin

In recent times the skin of the larger fishes such as sharks, are easy to use for preparing ladies shoes, hand bags, wallets, suitcases and belts.

13) Ornamental value of otoliths

The presence of exceptionally large otoliths finely rubbed and polished stones are usually affixed on silver rings and are used as jewelry.

14) For decoration or Aesthetic value

As a hobby some beautiful colored fishes are cultured in aquaria for the decoration of the houses e.g. Gold fish, Angel fish etc.

15) Biological control

Several species of fish seed on insect larva and are used to control mosquitoes, which spread diseases such as malaria.

16) Medicinal uses

Several fish species are used in medicine to cure the various diseases such as shark to extract the insulin. Otoliths used in convalescent children suffering from ricket and mucous of *Amphipnous cuchla* used in impotency.

17) Fish wafers and Fish biscuits

Certain fish flour is mixed with wheat flour and prepared the nutritious food as fish wafers, fish biscuits from the flour of fish.

15

Sexual Dimorphism in Fishes

The embryogenesis or embryonic development of a sexually reproducing multicellular animal is prefaced by the gametogenesis as ovum and sperms.

In some animals both kinds of gametes are produced by a single individual which is called monoecious or hermaphrodite.

But in majority of animals both kinds of gametes are produced by different individuals to an animal species are called dioecious.

The kind of sex cells as spermatozoa and ova; and those organs which are immediately concerned with their production i.e. male gonads called as testes and female gonads as ovaries form the primary sexual characters of each sex.

Besides these primary sexual characters the male and female sexes differ from each other in many somatic characters, which are called secondary sexual characters.

Such a phenomenon of morphological differentiation or specialization in secondary sexual characters between male and female individuals of a species is called sexual dimorphism. Sexes are generally separate in fishes and only a few species are hermaphrodite.

Sexual dimorphism

Females are generally larger in size and more numerous than the males. Several species show well marked structural differences in the two sexes. These are of two types such as:

1) Some species show morphological peculiarities which facilitate fertilization of ova, as copulatory organ in the male.
2) Structural peculiarities are not related to copulation.

1) Coputatory organ in the male: Fertilization is internal in chondrichthys and males possess claspers for transferring the sperms in to the body of the female. A clasper is a pair of elongated rod – like structures each having a cartilaginous skeleton and a groove along its whole length. During copulation the claspers are introduced into the cloacal aperture of the female which is found in male dog fish or scoliodon. Hence male scoliodon is separate from female scoliodon, which forms claspers as copulatery organs.

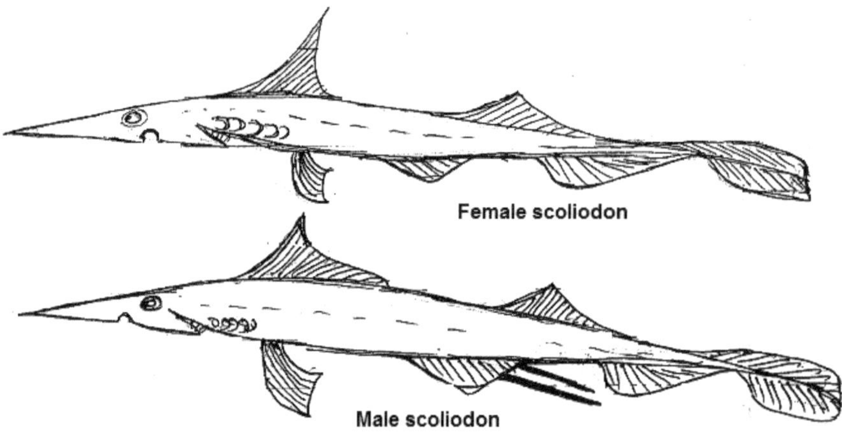

Fig. 15.1 : Male and female scoliodon

In holocephali, in chimaera fish, a club shaped frontal clasper is also present on the head of the male fish. It is believed that the male uses its frontal clasper to retain its hold on the female while curling its body round the female during copulation.

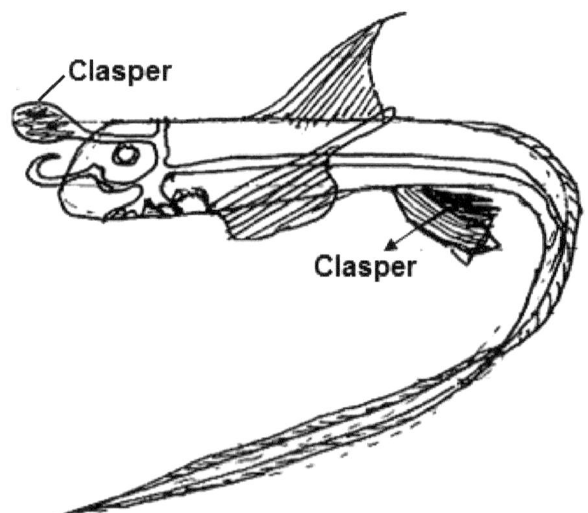

Fig. 15.2 : Male Chimaera

In teleostemii, fertilization is generally external in water. But in a few species a copulatory organs is present in the male, and internal fertilization takes place. A very simple condition is seen in the catfish Mystus seenghala in which a conical genital papilla is present in the male.

In gambusia affinis the vas deferens extends as a tube up to the end of the anterior rays of the anal fin, which are modified to form an elongated tubular or grooved copulatory organ called as gonopodium. This gonopodium is present in male guppy and absent in female fish. Female gambusia is large than male gambusia fish. No any male is of similar colour, mass variation in colour in male fishes found. Male fish has various colors of body hence it is called as rainbow fish as shown below :

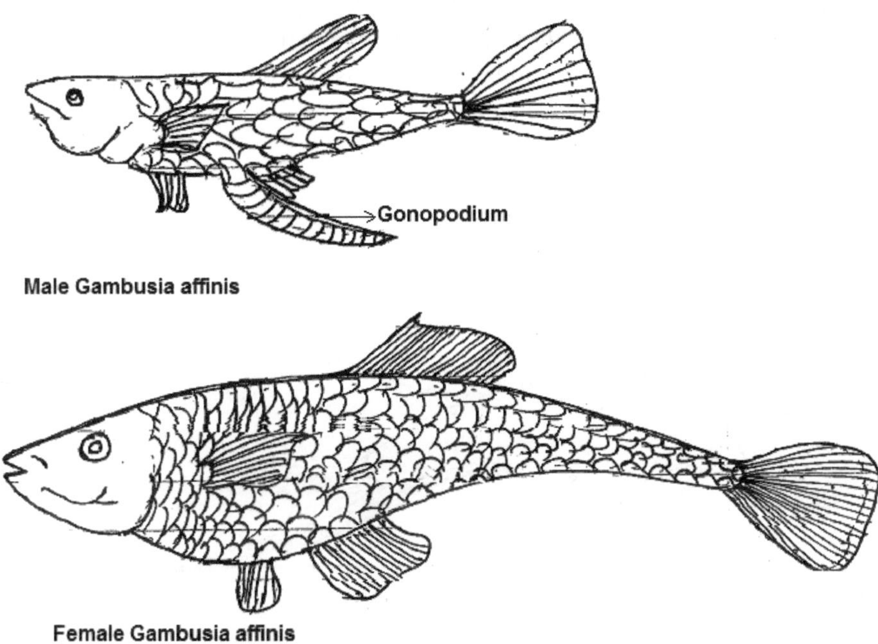

Fig. 15.3 : Male and Female gambusia affinis.

In the cyprinodont fish xenodexia the pectoral fins are modified and are used for holding the gonopodium in position for insertion into the oviduct of the female fish.

2) Structures not related with copulation : Well marked structural differences are seen in the two sexes in some fishes, especially during the breeding season, and these are not related to copulation. In most teleosts the female is larger in size than the male. A large fleshy hump is seen on the head of the male in Lion head cichlid fish.

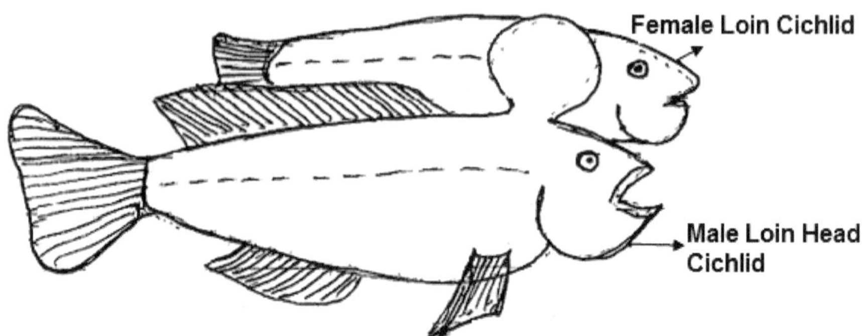

Fig. 15.4 : Lion head cichlid male and female fishes

In the swordtail fish is xiphophorus hallari the lower lobe of the caudal fin is elongated in male fish so as to form a blade like structures. Which is absent in female fish. In the male fish Caldas peduncle is broad both tubes of caudal fin is painted but the lower lobe is elongated and drown into a sward like.

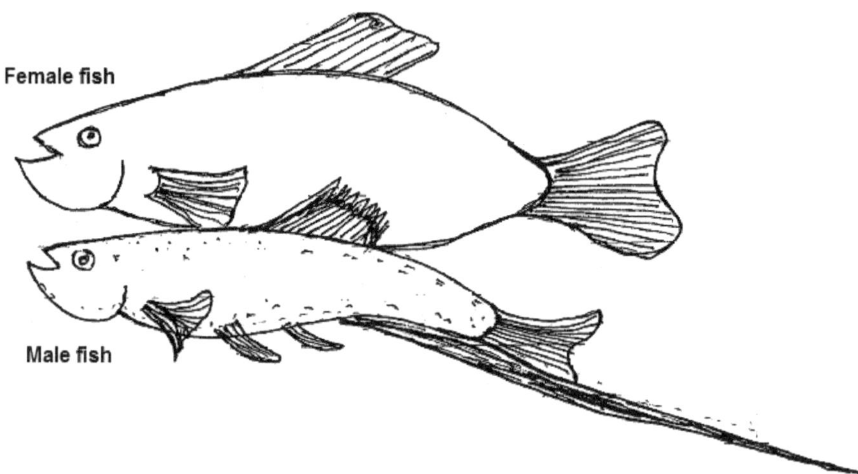

Fig. 15.5 : Male and Female Xiphophorus halleri fishes

Hippocampus is commonly known as sea horse it possesses the sexual dimorphism. Male sea horse possesses a brood pouch. Depending on this brood pouch we can identified the male Hippocampus fish in this brood pouch female lay the eggs to incubate the eggs and for shelter of young ones.

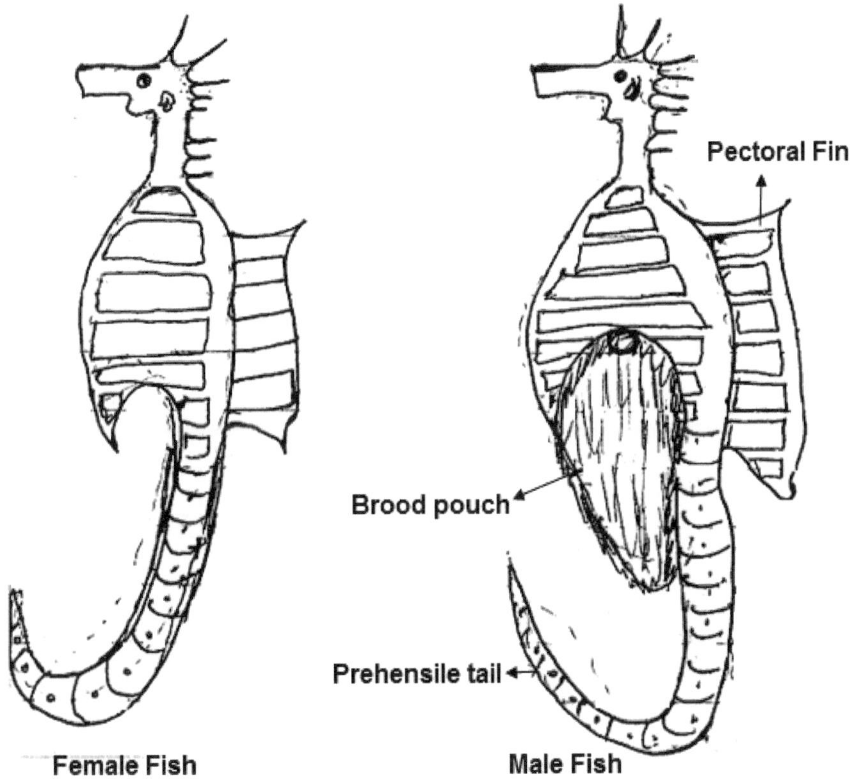

Pectoral Fin

Brood pouch

Prehensile tail

Female Fish

Male Fish

Fig. 15.6 : Female and male Hippocampous fish

The dorsal fin of the mate *mollenesia* is enlarged to form a sail like structure and has brilliantly coloured ocelli on it which is absent in female mollenesia fish.

The male bowfin amia has a large black spot at the base of the coudal fin and which is absent in female Amia fish.

Fig. 15.7 : Male Amia fish

16

Integrated Fish Farming

- Fish cum-poultry farming.
- Fish cum-duck farming.
- Fish cum-pig farming.
- Fish cum-paddy farming.

A) Fish Cum-poultry farming

Inland fish culture in conjunction with poultry rearing is a compatible business and can provide a readymade fertilizer for fish pond. In a larger sense fish and poultry are mutually complementary the former provides fish meal a protein rich poultry feed and the later is droppings which acts as fertilizers for plankton production to serve as natural proteinous fish feed. Fish meal is recognized as perhaps the choicest poultry feed. In an integrated fish cum-poultry farm, birds have to be housed in deep litter of battery system in which the pen flora is covered with hay. The floor of the pen is covered with chapped straw dry leaves, hay broken up groundnut shells maize straw and saw dust. Each bird should have about 0.3 m^2 space to live on the deep litter flooring bird droppings falling on the litter react with the flooring organic material. The litter has to be periodically churned for aeration and maintenance of hygienic conditions.

25-30 birds/hen are adequate to produce tones of fertilizer per year. Approximately 60 kg/ha/year of deep litter manure is applied in fish culture pond. Its application should be deferred and it should induce excess algal bloom in the water. Approximately 500-600 poultry birds are enough to sub serve the needs of one hectare of fish culture pond. Normally deep litter poultry pen fertilizer is obtained at Rs.-20 per quintal meat or eggs type birds are to be chosen. Rhode Island or Leghorn fowls would ordinary speaking suffice. The fish culture is done in ordinary manners but fertilization and feeding is not done except putting the poultry dropping. The economics of one hectare fish cum-poultry farm is given as under:

Table 16.2 : Income and expenditure on fish-cum-poultry culture

Items	Quantity	Rate in Rs.	Total
Expenses :			
1) Lease money of pond (hect.)	one	3000	3000.00
2) Depreciation of pond repairs and poultry shed construction (10% the Rs.44000)	—	—	4400.00
3) Electricity and water charges of Ponds (L.S.)	—	—	3000.00
4) Cost of fish seed	10000	50/1000	500.00
5) Eight weeks chicks (10% marrying of mortality)	550	7/bird	3850.00
6) Fowl feed (kg)	18350	2.00	36700.00
7) Medicines Cost (L.S.)	—	—	500.00
8) Netting charges (L.S.)			1000.00
9) Labour Charges (man days)	530	30.00	15900.00
10) Contingency (L.S.)	—	—	1800.00
11) Interest on the capital Investment	—	—	500.00
		Total	75000.00
Return :			
1) Sale fish (kg)	3500	16.00/kg	56000.00
2) Sale of eggs (Numbers)	70000	70.00/100	49000.00
3) Sale of chicks (live wt.kg)	1250	16.00/kg	20000.00
		Total	**125000**
Net profit :- Return – Expenses = 125000 – 75000 = 5000			

During fish cum-poultry farming generally above 50% profit is earn on the investment.

B) Fish Cum-duck farming

The fish cum-duck farming system not only results in more economic benefit to the farmer also the fish and ducks are benefitted by their co-existence. The dropping of ducks acts as substitute fish feed and pond fertilizers which account for so percent of the total input cost in fish culture. The ducks feed on such organisms from the pond such as larvae of aquatic insects, tadpoles, mollusks aquatic weeds etc., which do not form the food of stoked fish. The kind of duck to be raised must be chosen with care since all the domesticated races are not productive and hardy enough for keeping in open ponds in all weathers. Local variety, "Indian runner" is hardy and found to be most suitable for this purpose.

It has been found that 200-300 ducks are sufficient to produce manure adequate to fertilize hectare of water area under fish culture. Two to four months ducklings are kept on the pond after providing them necessary prophylactic medicines as a safeguard against epidemics compared to chicken, ducks are subjected to relative few diseases. A sick bird becomes restless; eyes loose brightness and discharge sticky fluid from eyes and nostril increases. The sound of sneezing and coughing from duck is an alarm for coming disease.

The duck dropping act as excellent fertilizer and the dobbing of ducks in the pond water in search of feed releases nutrients from the soil which enhance the ponds biological productivity and consequently increases the fish yield. This integrated farming has considerable scope in different states of our country in recent days.

The ducks do not need much elaborate house since most of them during day time remain in the pond. A house on the farm which is not being utilized for any other purpose can be used for keeping ducks. A small low cost shelter for ducks can also be constructed on pond embankments by using the splits of bamboo or any other cheap wood. A floating duck house can also be constructed on the water surface by using the empty barrels as floats. The duck house should be well ventilated and to keep the duck house clean and dry as for as possible. The over crow dings of ducks in the night shelter should be avoided and about 0.3 to 0.5 meter of floor space is enough for night shelter for each duck the natural food from the ponds are not enough for these proper growth therefore the supplementary feed a mixture of balanced poultry feed and good quality rice bran in the ratio of 1:2@ 100 grams/bird/day is given. The feeding is done in the morning and evening in the duck house in the shallow plates. Efforts should be made to keep duck dry and be maintained in proper hygienic condition. Duck house should be washed and die infect at least once a week.

The ducks start laying eggs after attaining the age of 24 weeks and continue to lay eggs for two years. The ducks lay only at night. It is advisable to keep some straw or hay in the corners of the ducks house for egg laying. The eggs are collected every morning after the ducks are let out of duck house. The local variety of ducks “Indian runner” lay 180 – 200 eggs in one year. Duck attains the weight of 2-2.5 kg in the period of two years.

The fish farming is done as usual without fertilization and supplementary feeding. The average annual fish 3500 kg is produced.

The rearing period is generally kept as one year and fish yield ranging from 3500-4000 kg/ha/year and 200-600 kg/ha/year, can be expected with six species. Catla, rohu, mrigal, silver carp, grass carp, and common carp and three species of Indian major carps respectively. The ducks are sold after completion of three years in view of decreased egg laying capacity in fishe-cum-duck rearing. The variable cost and return fluctuation of fish-cum-duck culture are framed in different in culture system on different managers.

Table 16.2 : Income and expenditure on fish-cum-duck culture

Items	Quantity	Rate in Rs.	Total
Expenses :			
1) Lease money of the pond (hect)	01.00	3000.00	3000.00
2) Annual depreciation of pond repairs and duck house (10% of 44000Rs)	—	—	4400.00
3) Pond preparation water charges (L.S)	—	—	1000.00
4) Cost of ducklings (Nor)	200	12	2400.00
5) Cost of fish seed(Nos1000)	10000	50	500.00
6) Duck feed {kg)	2300	2	4600.00
7) Medicine charges (L.S.)	—	—	500.00
8) Netting charges (L.S.)	—	—	1000.00
9) Labour charges (man days)	530	30	15900.00
10) Contingency	—	—	1200.00
11) Interest of the capital investment (12%)	—	—	5500.00
		Total	40000.00
Returns :			
1) Sale of fish (kg)	3500	16.00/kg	56000.00
2) Sale of duck eggs(Numbers)	18000	1.20 each	21600.00
3) Sale of duck meet (kg)	500	16.00/kg	8000.00
		Total	85600.00
Net profit = Returns – Expenses = Rs 85600-Rs 40000 = Rs. 45600			

During fish cum duck farming generally above 50% profit occurs.

C) Fish Cum-pig farming

The raising of pigs can be fruitfully blended with fish culture by setting animal housing units on the pond embankment in such a way that the waste and washing of house are drained into the fish pond. In this integrated culture system, the pig casting is used in pond. In this supplementary fertilization and feeding are not required for fish culture on an average 3500 kg fish per hectare pond is produced. The tradition composite fish culture in the pond is done but no supplementary manure and feeding except pig dung is added to the fish pond. On the other hand growth of pig depends on many factors including freed and strain of pigs.

The exotic land race varieties of pigs are most suitable as they have faster growth and least occurrence of diseases. Local Indian variety through very much resistant to diseases is slow growing. About two months old piglets are kept for fattening. Usually such pigs attain slaughter maturity (60-70) within six months through intensive rearing. These two batches of piglets can be raised along with fish which are cultured for one year. It has been found that the excreta discharged by 30-40 pigs in a year time is adequate for fertilizing one hectare water area under fish culture. The pigs are not allowed to go out of the pigs sites and balanced diet pig mash @ 1 kg/pig/ day is given as feed. The pig mash contains 30% rice bran 15% rice polish, 30% wheat bran, 10% maize bran 10% ground nut cakes, 1% common salt and 4% fish meal. Rovimix (vitamin - A, B12, D) 20 gm per 100 kg feed is added to make it more nutritive. Grass, green cattle fodder and drinking water are also provided to pigs.

Good housing with adequate accommodation is co-operating all essential requirements of pigs must be provided to keep them in healthy condition. The pigs are raised under two systems (1) Open air system and (2) indoor system. A combination of the two is followed in fish-cum pig farming. A single row of the pig pens facing the pond is constructed on the pond embankment. The pig pens must have arrangement for enough air, sunlight, exercise and dunging space. The feeding and water troughs are also to be built. The gates are provided to be cemented with drainage to the pond. The pig sites can be constructed by using widely traditional and locally available material. Bamboos are most suitable for making low cost pig sites. The height of pig sites should not exceed 1-5 meters whereat the height of the walls separating the pen and surrounding the run should not exceed of meter. For healthy growth of pigs a living space 1 to 1.5 meter is provided for each pig. Although, pigs are hardy animals yet they may be infested with parasites. The proper hygienic conditions of pig sites would keep pig away from danger of parasites. Although, pigs are not water animals yet they like taking baths and hence there would be no hesitation in washing them. Pig-sites should be washed daily in the morning hours to drain out whole excreta into the fish ponds. Disinfectant must be used at least at weekly intervals. It is advisable to get all piglet vaccinated against swine fever before keeping them for fatting. After attainment of marketable size pigs are marketed. Partial marketing can also be done.

Due to abundance of natural food in integrated fish-cum-pig pond, some fish attain marketable size within a few months instead of a year. Partial harvesting of the table sized fish done. Final harvesting is done other 12 months of rearing.

The following benefits occur from fish cum-pig farming:

1) The fish utilize the feed spilled by pigs and their excreta the former being

very rich in nutrient for fish and the latter for water.

2) Pig dung acts as a substitute for pond fertilizer and supplementary fish feed up to a point. Therefore the cost of fish production is greatly reduced.

3) Not much of additional land is required for piggery operations.

4) Cattle fodder required for pigs and grass carp is groan on the pond embankments.

5) The fish pond provides much needed water for washing the pig sites and the pigs.

6) It results in high production of animal protein per unit area.

7) It ensures high profit through less investment.

8) The mud which gets accumulated at the pond bottom is useable as a manure for agricultural and cattle fodder crops.

From the fish cum-pig farming near about 60% profit is found to the fish farmers.

D) Fish–Cum–paddy farming/Fish culture in rice fields

The paddy-cum-fish farming or culture is an ancient practice for the Far East. It has considerable potential for fish cultivation. The rural agricultural economy is largely strengthened by this practice. The advantages of this farming are as follows:

1) The rice and fish are cultivated either simultaneous or alternately in the same water mass.

2) Besides the agricultural produce of rice from the fields, fish is obtained at low cost to offer feed of animal protein in addition.

3) Fish cultivation brings about a decided improvement in paddy crop by way of exercising an effective control of unwanted weeds.

Raising fishes in paddy fields is one of the most rational methods of exploitation of water covered areas of agricultural land. In India although over 6 million hectare is under rice cultivation, the fish culture practice is restricted to only 0.03% and has yet to find a wide application.

The various hydrobiological conditions of rice field differ from of shallow water bodies like the ponds. Those influencing the culture of fishes are as follows.

1) Water in rice field has a limited depth, varying from 5 to 25 cm. depending upon the type of rice being cultivated.

2) Owing to the little depth the water in paddy fields warms up earlier and

the fish under cultivation should be able to adapt to higher temp up to 30°C or more.

3) The dissolved oxygen contents are very low, particularly during the night when plants consume oxygen and decomposition of different organic substances continues.
4) The turbidity is greater.
5) Acidity (pts) is more.
6) Other physico-chemical conditions are determined by the type of soil.
7) The feed stocks are not high and the differences depending upon agronomical practices, and conditions characteristics of different rice growing areas are found.

Not all fishes are suitable for paddy cum fish culture but depending upon ecological fish species suited for fish culture in paddy fields are those :

1) Fish can adapt to shallow waters necessary for paddy crops.
2) Fish can tolerate high temp and can support low dissolved oxygen content so characteristics of paddy fields.
3) Fish has a fast growth rate so as to attain the marketable size in short period.

The paddy fields need some management to increase their utility as fish ponds:

1) Some device to drain the field or to supply water to it.
2) Some device to keep the water at the desired level.
3) Some means to control the entry and exist of water in order to prevent the cultivated species from escaping and the unwanted wild fish from entering the paddy field.
4) Some device to let out water, if flooding occur.
5) Some means to provide shelter to the cultivated fish in moments of distress.
6) Some device to maintain a continuous flow of water in and out to the field.

Simultaneous and alternate paddy and fish cultivation:

Both simultaneous and alternate cultivation of paddy and fish have some advantages and disadvantages. The advantages of simultaneous cultivation are as follows.

1) No additional cost for fish production.

2) Rice production is increased by 5 to 15 % due to fish cultivation

Fish destroy weeds cause tillering and mineral enrichment by their digging activity and are indirectly responsible for the fertilization of soil by their excrement.

The disadvantages of simultaneous cultivation include:

1) Deep water condition necessary for better fish growth is not available in these farms.

2) Limitation on use of herbicides and insecticides for good rice production but fear of their harmful effect on fish.

On the other hand, alternate cultivation i.e. rotation of paddy and fish cultivation one after the other, has its own advantages which may be as mentioned below:

1) Water depth is no limitation either to paddy or to fish water is kept shallow until rice harvest then the depth is increased to suit fish raising.

2) The same field may be better managed from the point of view of paddy or fish production without interference to either.

Various techniques are employed for fish culture in paddy fields depending upon the climate, local conditions, species of fish available and the variety of paddy cultivated. Every paddy field produces some quantity of wild fish that enter as fry. The cultivation of paddy is the primary purpose to the farmer hence the fish culture is to be adapted to the schedule of paddy cultivation.

The rice fields in India differ in size from place to place. The area larger than 5 hector is inconvenient for fish culture. The rice plots of 25x10 m are considered ideal for paddy cum fish culture. Each plot is bordered on all the sides by bunds of approximately 0.3 m height and 0.3 m in width paddy fields to be used for fish culture are provided with strong barrier wall to prevent drainage of water to convert the field into pond of desired depth. Each plot is dug along its middle to form a long central canal of 0.5 x 0.75 m width and depth respectively. This canal connects with the main supply through the sluice.

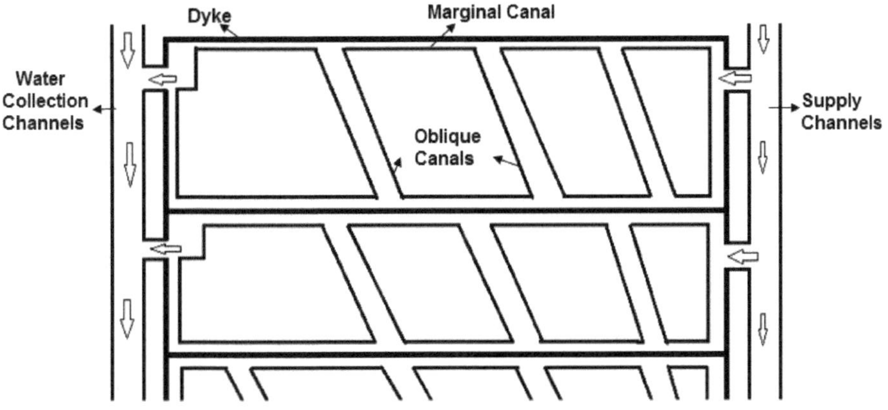

Fig. 16.1. Layout of model fish-cum-paddy fields

Cultivable Fishes

The suitability of fish depends upon their ability to withstand conditions of paddy fields. The fishes selected for cultivation should have faster rate of growth species such as catla, rohu, mrigala, cyprinus carpio, tillapia, anabus, *Clarius batrachus*, lates species, and channa species are widely cultured in rice field.

Experiments conducted on culture of channa striatus in paddy fields have shown 7-13% increase in the yield of paddy and an average production of 112 kg/ha of fish. Experiments conducted with tillapia, cyprinus carpio have also provided encouraging results. During the recent years high yielding varieties of paddy are being cultivated.

The air breathing fishes are considered good because of their accessory breathing organs which helps them to breathe when the fields gets dry and the water level in capture channel goes down. These fishes can survive well in shallow and turbid water of the rice field.

Fish cultivation in paddy fields differ from country to country from place to place depending upon the topography of land availability of water and the type of rice and fish to be cultured as:

1) Culture together with rice.
2) Culture in rotation with the rice.

It is essential to make a well planned survey of prevailing conditions before attempting fish culture in paddy fields on large scale.

The paddy cultivation has to be initiated in the month of April and May. After preparing the paddy may be selected for direct sowing after the first shower of monsoon rain. The fish seed @ 10000 per hectare is stocked in the trenches after filling with in monsoon season. Usual practice of fish culture is adopted. The Kharif crop is harvested by November and December months production of paddy ranges 4-5 tones/ha. The second crop of paddy is sown in the month of January normally with high yielding varieties such as Jaya, Ratna, Rust etc. The end of Rabi crop of paddy would be ready for harvesting by the month of May to June. At the same time the fish crop is also ready for harvesting. The fish production on an average is 3000 kg/ha/year.

17

Bio-Chemical Composition of Raw Fish Value

The per capita consumption of fish has been 3.2kg on an average up to 1992 as against estimated requirement of 11.0 kg pisciculture has the potentiality of popularity. Fish is rightly considered as the "Poor man's diet". It is almost zero carbohydrate food, good for diabetics and other such patients. Fish is a rich source of protein, vitamins and minerals. A special feature of fish flesh food is content of vitamin B12, which is almost absent in plant food and also a good source of calcium and vitamin A. Fish also contains polyunsaturated fatty acids which are known to provide protection against cardio-vascular diseases. This has got advantages over the other meat food.

Planktophagus fishes have higher protein contents then do fish with other feeding habits.

When flesh of fish is put to an analysis the following major components are observed.

Component	Quantity
Moisture/water	53.89%
Protein	20%
Carbohydrate	Negligible (only as Glycogen in liver)
Fat	Variable
Minerals	1-2%

Besides above the enzymes, pigments and vitamins are also the important constituents.

1. Proteins

Herbivorous or planktophagus fish have higher protein content than the fish with other feeding habits. This ratio is correlated with the gain in weight in some fish muscle is a rich source of protein -15-25%. About 95% of the protein nitrogen can be extracted by sodium chloride solution. This ratio is correlated with the gain in weight in some fishes. Such comparisons are possible among the individuals belonging to the same species.

Several factors are known to affect or influence the rate of growth in fishes. These are temperature, photoperiod, quality and quantity of food available, dissolved oxygen. Ammonia, salinity, age and the state of maturity of the fish, inter-specific and intra-specific competition among the individuals and crowding and disease etc.

The fractionation of the extracted protein gives mainly three types of proteins as Albumin-16-22%, contractile proteins-75% as myosin, Actin, Actomyosin and tropomyosin and stroma proteins-3%. The non-protein nitrogen forms 7.5-17.1 % of the total nitrogen.

Enzymes: Fish muscle contains appreciable quantities of proteolytic enzyme, Amylases, lipases, thiaminase, adenosine triphosphatase, Choline esterase and other nuclease, glycogenase have been reported. The antithiamine factor 'Thiaminase' has been found mainly in fresh water fishes.

2. Fats

The esters of fatty acids and glycerol constitute the storage fats of fishes. Although fat is distributed in all tissues when fat is present in extraordinary amounts such fats are called depot fats. The depot fats are found in muscle, head tissue, roe milt, liver, skeletal tissue, viscera etc. Brain shows the highest concentration of fat and heart the lowest.

Depending upon the fat content the fish may be classifies into three groups:

1. Oily or fatty- fat more than 8%
2. Average fatty- fat content between 1%-8%
3. Lean-Fat content is less than 1%

PUFA : There is high content of Polyunsaturated Fatty Acids (PUFA) in fishes. The PUFA are divided into three families namely oleic, linoleic and linolenic. The omega (ω) system is used to nomenclature of families. Fresh water fish appear to have higher levels of omega (ω^6) six fatty acids than marine fish.

3. Vitamins

Fish flesh are good source of vitamin A, D, E and B-complex mainly thiamine, riboflavin and nicotinic acid. Fish liver oils constitute the principal natural source of vita-A, oils of some species are rich in vitamin D. Vita-D is quite abundant in liver oils of cod and halibut. In general the less the oil content of the liver the greater is the vitamin-D content.

4. Minerals

Bulk of the minerals are concentrated in hard parts such as scales, bones, otoliths etc. Various minerals are found these are calcium, magnesium, potassium phosphorous, sulphur, iron etc. Phosphorous is the important constituent of Phospholipids and nucleotides. Potassium predominates over sodium in the tissues of both marine and freshwater fishes.

5. Carbohydrates

Fish flesh contains negligible quantities of carbohydrates. Glycogen is present in living fish and is rapidly converted to lactic acid after death.

18

Role of Remote Sensing in Fishery

The role of remote sensing in fishery science is one of the established fact. But today remote sensing is used generally everywhere. Success of fishing depends upon capacity for fish detection. Traditional methods like sighting, listening, smelling, locating phosphorescent patches, blue or yellow patches and oily patches in water are absolute now. Application of acoustic devices such as echo sounders, sonar's started in 1930. Satellite remote sensing technologies have been applied recently since 1980.

In India acoustic fish finder was used in 1946. However its application has remained insignificant. No acoustic survey has been carried out either in the Arabian sea or in the Bay of Bengal. In the Indian Ocean some headway was made in 1977 and there in 1982.

Satellite remote sensing is restricted to the surface of the ocean at present. The Satellite cannot use its sensors to penetrate the column of water. The two important ocean parameters in remote sensing of fish stocks are the colour and surface temperature. Colour is used to assess and map chlorophyll productivity and through fish productivity. Temperature on the other hand helps locating thermal fronts and upwelling Zones. There by assessing nutrient abundance as well as fish abundance.

In the Indian waters, air craft remote sensing has been used for locating shoots of pelagic fish like oil sardines and mackerels in the Arabian sea using colour and bioluminescence only. The signals help distinguish shoals by species and estimate the weight of the schools. Ocean colour map also help in locating tuna schools, which prefer clear oceanic waters as against the turbid.

Coastal waters satellite remote sensing technique is yet to establish itself in Indian waters. In India generally light and fish finder as Ecosounder and sonar are used on a large scale from ancient years. Light - fishes show phototaxis when they are stimulated by an artificial light or natural bright light, both above water and under water light produce this effects. The attraction of fish in great numbers by light has thus found application in fishery for capture of fish on commercial scale. It is now long that this method is being used. This method is particularly useful for pelagic fishes. By introducing this method the catches could be greatly increased the transparency of water is very necessary for best results. Intensity of light and color of lights also useful in fishery science the red light is useless being totally ineffective where as yellow light in combination with white light is found to be very effective, where as blue light is found to be good for attracting fingerlings.

Fish Finder

Fish finding in fish location by means of hydroacoustic device. A hydroacoustic device such as sonar or ecosounder is capable of determining the location of navigational objects that are anywhere in vertical or horizontal place inside water. The principle of action is simple. The hydroaucostic instrument emits sound waves. These waves travel in all directions in plane. Those waves which are incident upon underwater objects that come in their path get reflected. The reflected waves retrace the path and upon reaching the instrument serve as an echo signal for the instrument. This signal is received by the instrument. The nature of the object and its distance from the instrument are then worked out by an automatic recorder. The distance determined on the basis of the time interval that passed between the moment of emission of the sound waves and the moment of their reception. The nature of the object is determined easily, because different object have different reflecting properties. Single or shoal of fishes also posses the reflecting properties, fish finding with hydroacostic devices become possible and is currently used with success in commercial fishing. The main three purposes for fish finding are as follows:

1) To study some aspects of fish biology such as distribution of fish, movement of fish in horizontal and vertical planes, shoaling and migratory movements, behavioural response of fish in vicinity of a fishing net.

2) To study and reveal approaching shoals with respect to the composition, density and size of the shoals and the direction and speed of their movement.

3) To study fish ecology with respect to bottom configuration, distribution of underwater obstacles, submerged and bottom vegetation and so forth.

Fish behaves as a navigational object in reflecting sound waves incident upon it in the form of echo-signals for the hydroacoustic device. The reflecting property of fish is very different from that of other under water objects. Fish in shoal produce multiple reflections as compared to single reflections from solitary fish.

A fish finder is a hydroacoustic instrument specially designed for use in fisheries. It is now becoming an indispensable tool for promotion of commercial fisheries. The efficiency of a fish finder is judged by its range of action. The range of action may be determined with respect to both the sensitivity and the geometric image produced of the acoustic field.

A fish finder consists of two parts a transmitter vibrator and a receiving recorder. The transmitting vibrator emits ultrasonic pulses. In the case of a fish finder working on horizontal plane, the reflected ultrasonic wave or echo signal is converted by the receiver into audio-signal if the echo-signal is to be heard by the operator for interpretation.

There are three principle models of fish finder as sonar, echo sounder and fish magnifier. The sonar and echo sounder work on same principle and in similar manner. The fish magnifier is a special design which is used for investigation of a very small section in the vertical plane or depth but which furnishes much more detailed picture of the facts.

Echo sounder

Echo sounder is smaller and more handy than sonar. Sonar has very complex electrical circuit in its design. Echo sounders are more widely used on small and medium size fishing operations and for scouting and investigation ships.

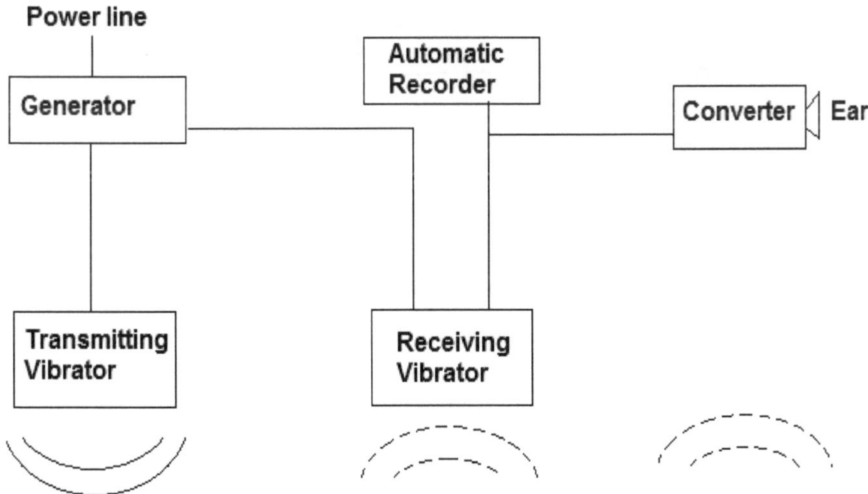

Fig. 18.1 : Ray diagram illustrating function of Echo sounder.

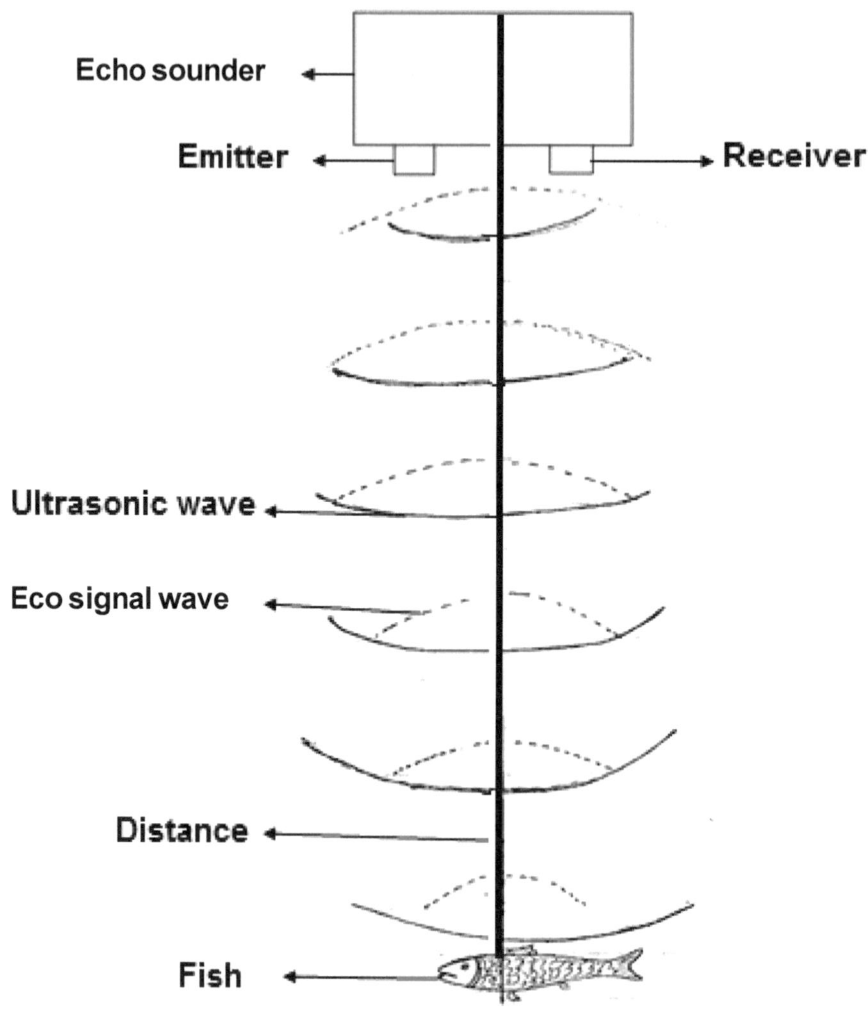

Fig. 18.2 : Principle of fish finding (echo sounder, sonar)

Some of the echo sounders are capable of recording distribution of diffuse fish concentrations at a depth of 380 m. The recording of signal or shoals takes place in the form of echograms if automatic recording is done. Slow moving solitary fish produces a crescent shape on the echogram. These various instruments are used in fishery science.

Sonar

The main use of sonar is to locate fish grounds and directs the vessel towards the school of fish. Constantly keeping touch with their movement, it is a long range fish finder capable of steering the transducer beam in all angles for search purposes. During search operation the behavior of sonar is exactly similar to radar. Narrow beam is used for deciding the correct bearing of the fish and wide beam is used while searching and approaching the fish school. Depending upon the range selected the pulse length is controlled.

Since the maximum detecting distance of fish schools with sonar is limited by refraction of sound beam due to thermocline, pressure and salinity, attenuation of sound with distance and scattering on the fish body, sonar could be used only to detect fish schools at low speed of the vessels within a range of 700 to 1000 meters. The echogram chart gives, not only the distance between the vessel and the school of fish but also additional information such as abundance and the type of fish. The scope gives both distance and horizontal picture of the school of fish which will be very useful for mid water trawling. New developments on the present day sonar are digital steady picture display and three dimensional displays.

19

Indian Exclusive Economic Zone (EEZ)

The concept of Exclusive Economic Zone, popularly called as EEZ was born out of a conflict on fishing rights among coastal nations. Some nations found their waters being exploited by foreign countries equipped with better and more powerful fishing gear and craft more sophisticated navigational and fish detection systems than their own. In 1950 several such nations unilaterally established fishing limits in their coastal waters for exclusive fishing. In the beginning the limit was just twelve mile. Again in 1970 some more countries are added and then they unilaterally accepted the limit of 200 nautical miles.

The United Nations General Assembly convened during 1974-1982 conferences on the Third law of the sea. The United Nations convention on the law of the sea (UNCLOS) was the outcum of the deliberations after it was ratified by as many as 60 countries in 1994. According to the UNCLOS each coastal country shall have an EEZ of 200 nautical mile zone around the share with exclusive right to exploitation of marine resources. The UNCLOS also imposed certain responsibilities on them. Each Country shall have a duty to properly manage its resources. The resources not being fully tapped by a country other countries shall legally have the right to exploit such waters by use of trawlers. It was also stated that beyond the EEZ, the fisheries of the high seas shall be regulated by the Regional Fisheries Organizations. Thus straddling fish and migratory fish fisheries become the subject of not the coastal states but of the regional fisheries organizations. This agreement was reached in 1992 at the Rio-de-Janerio United Nations conference on Environment and Development (UNCED) which worked under the frame work of the UNCLOS. About 90% of the marine catch is taken in the EEZ many countries including India have yet to realize their full potential from their EEZ. A few countries have not accepted EEZ restriction like South Africa. A number of countries on other hand are almost optimally exploiting their EEZ such countries are New Zealand, Iceland and Canada. The EEZ is a common shared area among member countries of European common market (BCM)

The imposition of EEZ to 200 mile posed great problem to a number of ocean fishing countries such as Russia, Japan. The EEZ concept actually envisaged a stretch of water from erstwhile 12 mile to 200 mile. i.e. excluding the inshore coastal water fisheries up to 12 mile which remained with traditional and artisanal fisherman for exploration. For many countries including India new waters of EEZ that were not only too far out but deep also were difficult to explore for lack of proper gear and vessel and technical knowledge. Joint ventures with foreign countries having experience and expertise were the only solution for such countries though very expensive. The EEZ did change the fishing pattern. For some countries over exploitation occurred in the EEZ causing damage to the inshore fishery often to the extent of provoking unrest. For other countries exploration of EEZ has been rather disappointing.

The EEZ has pushed up fish price and fish poaching in others unprotected water. Quality is another consideration that comes in the way of selling and marketing fish to keep EEZ viable. Proper management is the only hope for the future of EEZ.

The Indian EEZ came into force through the 41st Amendment to the Article 298. The Indian Exclusive Economic Zone (EEZ) surrounding the peninsula and islands (6000 km coast) has an area of 2.20 million sq.km. Only inshore water up to 50 m depth is actually exploited presently while the offshore water is largely unexploited.

The Government of India declared 200 nautical miles from the coast line as nations EEZ in January 1977, which has changed the open access regime to a controlled fisheries management over a vast sea area earlier to this declaration, we had no control over exploitation of our offshore fisheries resources by the distant water. The declaration of EEZ has now provided us the opportunity for controlling fishing and of managing our fishable stocks.

The present production from the inshore water is 1.5 million tons per year accounting for 90% of the total coastal production. The offshore water production is only 0.2 million tons per year which is only 40% of the total production. The exploited inshore and offshore water comprise a zone with the 20 miles only of the Indian coast, the present catch of 2.3 million tons (1996) is just 70% of the estimated potential of 309 million tones. This is because fish is mostly confined to a depth of 50 meters. Beyond this depth the deep sea is negligibly exploited for lack of infrastructure. Deep sea fishing is holding great scope and is therefore a matter of immediate concern and attention.

The EEZ is almost 2/3 of the inland area of the country. At the present status of fishery of the EEZ it is clear that the offshore is most untapped. The estimated total potential of the EEZ is from 309 million tones to 4.5 million tones. The species composition of the potential resources of the EEZ include the tuna, anchovies, perches, ribbon fish, lizard fish, catfish, shark, squid and cattle fish, bivalves, prawns shrimps lobsters and crabs. Of these shrimps and prawns are the most exploited to the extent of overfishing while tuna, squid, cuttlefish and lobsters are the most untapped. To check the poaching unauthorized intrusion and smuggling into the Indian EEZ, patrol vessel, fast and well equipped, undertake surveillance, search and rescue work along the coastal area.

The fact that EEZ resources to the tune of about 3 million tones per year remain unexploited has been the basis for planning strategies for exploration and exploitation of the EEZ resources. The Govt. of India introduced several schemes from time to time.

The Govt. of India had been exhorting the private fishing sector to diversify fishing efforts to exploit fisheries in the deeper areas in the EEZ so as to increase the country foreign exchange. The exploratory survey of fishery survey of India (SSI) to diversify different types of fishing within the EEZ.

20

Commercial Fisheries Institution of India CIFE, CMFRI, CIFT

Though India is one of the major fish producing countries in the world, it has no history regarding technical and academic training in fisheries. India has to tackle simultaneously the long range problem of establishing a regular fisheries educational system as one of its fisheries policies for ensuring a reserve of large number trained persons. The short range problem of providing specialized training to the large number of in service personal, who were not basically qualified in all aspect of fisheries had been planned. Therefore, the government of India and the state Governments had put in considerable effort to solve problem of shortage of technical person as by organizing a number of fisheries training institutions in the country at different levels. A few universities are also providing training in fisheries leading to degrees in fisheries.

The following are the main fisheries training institutions which are giving special training in fisheries in India:

1. Central Institute of Fisheries Education(CIEF) - Versova Mumbai

This is a Deemed University. This institute was started in 1961. There is administrative control of ICAR (Indian Council of Agricultural Research) i.e. Government of India over this institute. The institute conducts two years post graduate Diploma course in Fisheries Science, equivalent to M.Sc. degrees of Universities. The Basic qualification for admission is B.Sc. in natural Sciences. The intake capacity is 40 students per year. The institute conduct postgraduate and doctoral degree programmes and also post-graduate diploma and certificate courses in Fisheries, Master of Science in Fisheries Management (M.Sc. FM), Master of Science in Inland Fisheries Administration and Management (M.Sc. IFAM), Diploma in Fisheries Science (D.F.Sc.), Certificate Course in Inland

Fisheries Development and Administration and Certificate course in Fisheries Extension are also imparted here. There is also one year certificate course in Inland Fisheries Operational Management. There is also Ph.D. course which normally takes three years in completion several training programmes in inland fisheries, marine fisheries, fisheries extension education, fisheries economic, fish processing, fish breeding and culture hatchery technology and fish seed production technology are conducted time to time for scientists and fisherman.

This institute is also organizing short term courses in subject like management of brackish water and freshwater cultures, management of prawn and carp hatchery, fish seed production, utilization of low priced fish and management of fishery co-operatives.

These short term courses last per two weeks only and any graduate or experienced person is admitted to the course, the institute has two fish farms in Andhra Pradesh and recently acquired at Powerkheda in Madhya Pradesh for practical demonstration and training purposes.

It has three sub-centers which conduct various courses as described below:

a) Inland Fisheries Training Centre, Barrackpore. A One year certificate Course in Inland Fisheries Development and Administration is conducted. Normally B.Sc. holders are admitted and the intake capacity is 40 persons per year.

b) Regional Training Centre for Inland Fisheries operatives, Agra. A Nine months certificate course in fish culture for Inland Fisheries is organized, In this Course matriculates are admitted.

c) Central Fisheries Extension Training Centre, Hyderabad. A Ten months certificate course in extension techniques is imparted. For its admission B.Sc. degree is needed.

2. Central Marine Fisheries Research Institute (CMFRI) Govt. of India, Cochin- Kerala

This institute was established in 1947 by the Ministry of Agriculture and Irrigation and later on brought under the Indian Council of Agricultural Research in 1967. The Head quarter is located at Cochin and its regional centre at Mandapam camp. There are as many as eleven research centers and twenty eight field centers situated along the east and west zone sea coasts. The institute is engaged in developing in suitable mariculture technology for fin-fish and shell fish production in order to transfer technology short term and long term trainings and post graduate programmes.

There are eight major divisions of research programmers which include- Physiology, Nutrition and pathology division, Fishery environment management division, Fisheries Resources Assessment division, Pelagic fisheries division, Demersal fisheries division, Molluscan fisheries division, Crustacean fisheries division, Fisheries economics and Extension division. The institute is affiliated to Cochin University of Science and Technology which conduct M.Sc. and Ph.D. programmers in mariculture.

3. Central Institute of Fisheries Technology (CIFT) Matsya Puri, Cochin- Kerala

The Institute was established in 1957 under the Department of Agriculture, Ministry of Food and Agriculture, Govt. of India. It was brought under the administrative control of Indian Council of Agricultural Research (ICAR) in October 1967. Though the Head Quarter was located at Cochin, its research centers are at Veraval –Gujarat, Burla- Orissa, Mumbai- Maharashtra, Kakinada-Andhra Pradesh, Panjim –Goa and Calicat –Kerala. Attempts are being made for the development of better qualities of crafts and gears. The institute conducts training programmes for the fishermen. Researches are carried on fish processing, Biochemistry, microbiology, engineering, electronics and statistics.

Many other agricultural universities in India have started fisheries faculties providing fisheries either as a special subject or run as full time courses. Aside from the highest level of educational institutions in fisheries, the state Government have also started fisheries schools and training centers for fishermen at lower level. Training centers for handling of the mechanized fishing boats in Gujarat, Maharashtra, Tamil Nadu and Kerala have proved very successful in demonstrating to the fishermen that, their catches can be considerably increased through mechanization.

In the states of Maharashtra and Tamil Nadu special attention has been paid to the education of fishermen with fishery knowledge is offered and children are provided with all facilities. They are taught about navigation carpenter, net making and repairing of fishing tackle. In Maharashtra there are ten schools for fishermen community and the number of students on the rolls has been increasing from year to year. One fisheries high schools with fisheries subject is also established at Satpati in Thane district.

Apart from the training in modern methods in fisheries the fishermen have kept a rich heritage and customs of fishing profession. As such they remained as a district class of professionals. Their vast experience in the sea will prove very useful in conversion of their skills to modern techniques fishing. They can easily take to deep sea fishing and advance fishing methods provided they get a short term training in those techniques.

So proper education is given in fisheries so that the trained will implement modern fishing techniques. This will enable the fishermen to a great extent to increase fish production in particular and develop the fish industry.

Thus establishments and functioning of all the important fisheries institutes are directly or indirectly controlled by Indian Council of Agricultural Research (ICAR) with its head Quarter at New Delhi- Krishi Bhavan, Dr. Rajendra Prasad Road New Delhi. The Fisheries Research Institute both fresh and marine water has been worked out fully in Indian waters.

Bibliography

Agarwal, N.K. 1996. Fish Reproduction. APH Publishing Corporation. New Delhi. 157pp.

Alikunhi, K.H. 1956. Fish culture techniques in India. In: Progress of fisheries development in India. Central Inland Fisheries Research Institute and Department of Fisheries, Orissa, India. pp 63-73.

Alikunhi, K.H. 1957. General notes on the breeding and young stages of *Etroplus suratensis*. In: Fish culture in India, *Farm Bulletin* No.20. ICAR, New Delhi. 144pp.

Allen, G.R.,1985. FAO Species Catalogue Vol. 6 Snappers of the World. An annotated and illustrated catalogue of lutjanid species known to date. FAO Fish. Synop., (125) Vol. 6:208 p.

Amesbury, S.S. and R.F. Myers, 1982. Guide to the Coastal Resources of Guam. Vol. 1. The Fishes. Agan, Guam, University of Guam Press, 141 p.

Ayyappan, S., S. Raizada and A.K. Reddy. 2001. Captive breeding and culture of new species of aquaculture. In: A.G. Ponniah, K.K. Lal and V.S. Basheer (Eds.). Captive Breeding for Aquaculture and Fish Germplasm Conservation. NBFGR-NATP Publication-3. Key paper No.1. National Bureau of Fish Genetic Resources, Lucknow, U.P., India.

Bell, L.J. and P.L. Colin, 1986. Mass spawning of *Caesio teres* (Pisces: Caesionidae) at Enewetak Atoll, Marshall Islands. *Environ.Biol.Fish.*, 15(1):69-74.

Bhatnagar, G.K. 1972. Maturity, fecundity, spawning season and certain aspects of Labeo fimbriatus (Bloch) of river Narmada near Hoshanagabad. *J. Inl. Fish.Soc. India.*, 4:26-37.

Chakrabarty, R.D. and D.S. Murty. 1972. Life history of Indian Major Carps, *Cirrhinus mrigala* (Ham.), *Catla catla* (Ham.) and *Labeo rohita*(Ham.). *J.Inland Fish. Soc. India.* 4:132:161.

Chatterjee, S.N and S.K.Sharma. 1994. Impact of thermal discharge on the distribution and reproduction of Indian major carps in Sarni Reservoir, *M.P. Fish.Chim.*, 14(7): 29-32.

Chaudhary, M., V. Kolakar and R.Chandra. 1982. Length-weight relationship and relative condition factor of four Indian major carps of river Brahmaputra, *Assam J. Inland Fisheries Society of India*, 14(2): 42-48.

Coleman, N. 1981. Australian sea fishes north of 30°S. Sydney and Auckland, Doubleday, 287 p.

Das, P. and Pandey, A.K. 1999. Endangered Fish Species: Measures for Rehabilitation and Conservation. *Fish. Chim.*, 19(6): 31-34.

Day, F., 1867. On some new or imperfectly known fishes of Madras. *Proc.Zool.Soc.Lon.*, 935-42

Desai, V.R.2003. Synopsis of Biological data on the Tor Mahseer, *Tor tor* (Hamilton,1822). FAO Fisheries synopsis No.158. FAO, Rome. 36pp.

FAO/UNDP. 1981. Conservation of the genetic resources of fish: problems and recommendations. Report of the expert consultation on the Genetic Resources of Fish. FAO Fish Tech.Paper 217:43pp.

FishBase (2015): http://www.fishbase.org/search.php.

Froese, R. and D. Pauly. Editors. 2006. Fish Base. World Wide Web electronic publication.

Goldman, K.J. (1997). Regulation of body temperature in the white shark, *Carcherodon carcharis*. *Journal of Comparative Physiology*. B Biochemical Systemic and Environmental Physiology 167(6):423–429.

Ghosh, S.K. and A.G. Ponniah. 2001. Freshwater Fish habitat Science and management in India. *Aquatic Ecosystem health and management Society*. 4: 367-380.

Helfman, G.; Collette, B. and Facey, D. (1997). The Diversity of Fishes (1st ed.). Wiley-Blackwell.

Jhingran, A.G.1990. Restoration of fisheries in the Ganga river system. Proposed Action Plan. Central Inland Capture Fisheries Research Institute. Barrackpore, India.

Kapoor, D., R. Dayal and A.G. Ponniah. 2000. Fish Biodiversity of India. National Bureau of Fish Genetic Resources, Lucknow, U.P.,India. 775pp.

Karekar, P.S and D.V. Bal. 1960. A study of the maturity and spawning of *Polydactylus indicus* (Shaw.). *Indian J.Fish.*, 7(1):147-165.

Khanna S S, Singh H R. 2009. A Text Book of Fish Biology And Fisheries. Narendra Publishing House, New Delhi.

Kulkarni, C.V. and S.N.Ogale. 1979. The present status of mahseer (fish) and artificial propagation of *Tor khudree* (Sykes). *J.Bomb. Nat. Hist.Soc.*, 75(3):651-60.

Lagler,K.F., Bardach,J.E., Miller,R.R., Passino, D.R.M. 1977. Ichthyology John Wiley & Sons, New York.

Masuda, H., C. Araga, and T. Yoshino, 1975. Coastal fishes of southern Japan. Tokyo, Tokai University Press, 382p.

Menon A.G.K. 1988. Conservation of icthyofauna of India, conservation and management of inland capture fish resources. Central Inland Capture Fisheries Research Institute, Barrackpore, India. Bulletin,57: 25-33.

Mina, M.V. 1991. Microevolution of Fishes: Evolutionary Aspects of Phenetic Diversity Oxonian Press Pvt. Ltd., New Delhi.

Munro, A.D., A.P. Scott and T.J.Lam. 1990. Reproductive Seasonality in Teleosts: Environmental Influences. CRC Press Inc., Boca Raton, Florida. pp. 109-124.

Munro, I.S.R., 1955. The Marine and Freshwater Fishes of Ceylon. Canberra, Department of External Affairs, 351 p.

Qasim. S.Z. 1966. Sex ratio in fish population as a function of sexual difference in growth rate. *Curr.Sci.*, 35:140-142.

Qasim, S.Z. 1973. An appraisal of the studies on maturation and spawning in marine teleosts from the Indian waters. *Indian J. Fish.*, 20(1):166-181.

Qureshi, T.A. and N.A. Qureshi 1983. Indian Fishes. (Classification of Indian teleosts). Published by Brij Brothers, Bhopal, M.P., pp. 224.

Rao, G. R. 1974 . Observation on the age and growth, maturity and fecundity of *Labeo fimbriatus* (Bloch) of the river Godavari. *Ind.J. Fish.*, 21(2):427-445.

Rao, G.R. and L.H.Rao. 1972. On the biology of *Labeo calbasu* (Ham.-Buch.) from river Godavari. *J.Inland Fish.Soc. India.*, 4:74-86.

Sahu A.K., S.K.Sahoo and S.Ayyappan. 2000. Seed production and hatchery management : Asian catfish, *Clarias batrachus*. *Fish. Chim.*, 19(10&11): 94-96.

Saidin, T. 1986. Induced spawning of *Clarias macrocephalus* (Gunther), In: J.L. Maclean, L.B.Dizon and L.V.Hosillos (Eds.). The First Asian Fisheries Forum. Asian Fisheries Society, Manila, Philippines. pp.683-686.

Sakhare V.B. Fish & Fisheries of Indian Reserviors, Astral International Pvt Ltd. India.

Sakhare V.B. 2009. Aquatic Ecology. Narendra Publishing House, New Delhi.

Sreenivasan, A. 1976. Fish Production and Population changes in some Reservoirs in Tamilnadu. *Ind. J. Fish.*, 23:124-152.

Sugunan, V.V. 2002. *Clarias gariepinus* (African Catfish) gravitates into Yamuna, Sutlej, Godavari Angst comes True. *Fish.Chim.* 22(7):50-52.

Talwar, P.K. and Jhingran, A.G. 1991. Inland Fishes of India & Adjacent Countries Vol.1&2 Oxford & IBH Publishing Company Pvt. Ltd., New Delhi.

Vishwanath, W. and K. S. Devi, 2005. A New Fish species of the genus *Garra* Hamilton-Buchanan (Cypriniformes:Cyprinidae) from Manipur, India. *J. Bombay Nat. Hist. Soc.* 102(1):86-88.

Yadav B.N. 2006. Fish and Fisheries. Daya Publishing House, N. Delhi.

Zairin, M.Jr., K.Furukawa and K.Aida. 1992. Induction of ovulation by HCG in the tropical walking catfish Clarias batrachus under 23-250 C. *Bull.Japan Soc.Sci.Fish.*, 59: 1681-1695.

NIPA Publications on Fisheries

S.No.	Title	Author	ISBN	Year
1	Advances in Harvest and Postharvest Technology of Fish	Nambudari,D.D.	9789381450093	2012
2	Ancestral Knowledge in Agri-Allied Science	Saha, Ratan Kumar	9789383305216	2014
3	Applied Bioinformatics Statistics and Economics in Fisheries Research	Roy, A.K.	9788189422868	2008
4	Applied Computational Biology and Statistics in Biotechnology and Bioinformatics (Set of 2 Vols.)	Roy, A.K.	9789380235929	2012
5	Emerging Technologies of the 21st Century	Roy, A.K.	9789383305339	2015
6	Evaluation and Impact Assessment of Technologies and Developmental Activities in Agriculture,Fisheries and Allied Fields	Roy, A.K.	9789380235400	2011
7	Fish Fermentation: Traditional to Modern Approaches	Baishya, Debabrat	9789380235103	2009
8	Freshwater Fish Parasites	Dash, Gadadhar	9789381450253	2012
9	Handbook on Freshwater Aquaculture	Singh, N.P. & B.Santhosh	9789383305544	2015
10	Physiology of Finfish and Shellfish	Samantaray, Kasturi	9789383305681	2015
11	Water Quality Modeling: Rivers,Streams and Estuaries	Manivanan, R.	9788189422936	2008